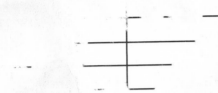

Technical Barriers to Agricultural Trade

Technical Barriers to Agricultural Trade

Jimmye S. Hillman

Westview Press
BOULDER • SAN FRANCISCO • OXFORD

This Westview softcover edition is printed on acid-free paper and bound in library-quality, coated covers that carry the highest rating of the National Association of State Textbook Administrators, in consultation with the Association of American Publishers and the Book Manufacturers' Institute.

Published in 1991 in the United States of America by Westview Press, Inc., 5500 Central Avenue, Boulder, Colorado 80301, and in the United Kingdom by Westview Press, 36 Lonsdale Road, Summertown, Oxford OX2 7EW

A CIP catalog record for this book is available from the Library of Congress.
ISBN 0-8133-8130-4

Printed and bound in the United States of America

The paper used in this publication meets the requirements of the American National Standard for Permanence of Paper for Printed Library Materials Z39.48-1984.

10   9   8   7   6   5   4   3   2   1

Contents

Tables and Figures

Figures

Foreword

Professor Hillman's *Technical Barriers to Agricultural Trade* is both timely and important. It is timely because the GATT negotiations are proceeding slowly in the agricultural arena partly due to a lack of agreement on technical barriers to agricultural trade. The book is important since it highlights many of the nontariff barriers which exist and concludes that they are significant impediments to international agricultural trade.

Professor Hillman has worked in international agricultural trade for many years and is a well-known scholar on the subject—one who has the ability to combine theory and practice. My first acquaintance with Professor Hillman involved a conference hosted in Tucson, Arizona, in April 1977. This, to my knowledge, was the first symposium held on international trade and agriculture. Up to that point, trade was generally a neglected aspect of agricultural policy. From these discussions evolved a book published by Westview Press entitled *International Trade and Agriculture: Theory and Policy*. Professor Hillman has added significantly to the knowledge of international agricultural trade since the time that volume was published. *Technical Barriers to Agricultural Trade* presents some of this added knowledge.

It is a significant textbook—at least in terms of supplementary readings for students interested in international economics. The readable and factual text makes this also an extremely useful book for policymakers. This book provides excellent background of the history of protectionism and the institutional framework within which protectionism takes place—Chapters 2 and 3 are especially strong in this regard. Chapter 4 presents an excellent categorization of the types of nontariff measures in agricultural trade, illustrating that there are many impediments to international trade in addition to commonly discussed tariff barriers. Chapter 5 focuses on the use of nontariff measures in developed countries, where the various nontariff measures and their relationship to GATT are discussed. In Chapters 6 and 7, Professor Hillman goes into great detail on specific nontariff barriers as related to trade in meat and livestock and poultry. These chapters clearly show the importance of nontariff barriers and the difficulty in resolving trade disputes in which nontariff barriers are important. In Chapter 8, Professor Hillman concludes with future prospects for technical trade barriers and agriculture and gives a concise overview of international trade negotiations and measures of protectionism. The role of GATT in agricultural trade liberalization and how nontariff barriers can be taken into account are throughly discussed in the concluding chapter. This chapter—along with Chapters 2 and 3—provides excellent material for policymakers.

Dr. Hillman's effort provides the basis of what will become an important and highly researched topic in the years to come. Several new nontariff barriers have surfaced since Professor Hillman completed this manuscript. For example, the Big Green Initiative in California, if passed, would limit the use of herbicides and pesticides in California agriculture. This could have a significant impact on California's ability to compete both nationally and internationally in agricultural production and trade. The "environment," which is essentially what the Big Green Initiative is all about, encapsulates the category of technical barriers to agricultural trade.

Trade and the environment will be a major focus in the 1990s. How one assesses the environmental consequences of agricultural production has important implications for policy and for the meaning of comparative advantage in agricultural trade. Changing the environmental indices on a worldwide scale changes the ranking of nations which have a comparative advantage in production and trade. *Technical Barriers to Agricultural Trade* provides the basis upon which environmental issues can be explicitly treated.

Andrew Schmitz
Robinson Chair Professor
University of California, Berkeley

Preface

Modes of agricultural protection have undergone a total transformation since the 1930s, a transformation which accelerated after World War II. Exporters of agricultural products once were faced by discreet customs duties that were often buttressed by sanitary, embargo or related legislation. Since World War II, they have increasingly become victimized by administrative and regulatory protective devices. In addition, onerous domestic agricultural policies have resulted in large-scale effective protection in many countries.

Examples of this neo-protectionistic trend abound in all countries, developed as well as developing, but are most common among the temperate-zone agricultural producers. New techniques to protect the agricultural sector from competition emerged in Europe after World War I when high tariffs and monetary controls proved ineffective in protecting domestic producers. The United States "upped the ante" by enacting an extensive series of legislative acts designed to directly improve agricultural prices and incomes. An integral part of this legislation was the now-famous Section 22 of the Agricultural Adjustment Act (AAA) of 1933. Other countries followed suit. It was only a matter of time until legislation everywhere would be designed similar to the comprehensive programs of the United States. Such was the Common Agricultural Policy (CAP) of the European Community which resulted from the 1958 Treaty of Rome. Japan's agricultural and trade policies became a major concern after that country emerged as an economic superpower.

Even though technical and nontariff trade barriers had been on the scene for many years, they proliferated noticeably after the Kennedy Round of tariff negotiations in 1962-67. Countries became more adept at designing administrative and regulatory measures to protect their agricultural sectors under the guise of new rationales such as food self-sufficiency, farm price and income stability, and rural development.

Moreover, the success in the GATT of lowering the customs duties and other direct taxes on trade now shifted the burden of protection to more subtle forms. The Tokyo Round of GATT negotiations did little to alter agricultural protectionism; in fact, many countries became more insistent on maintaining the *status quo* with farm production and distribution policies, even to the extent of massive export subsidization of farm products.

After two decades of little progress on the nontariff barrier and agricultural policy problems, the United States has insisted that the focus of the current Uruguay Round of GATT negotiations be specifically on these problems. It is too early to predict the ultimate and long-range fallout of the Uruguay Round, but my prognosis after the Houston Summit is that there will be more talks and less action toward removal of agricultural trade distortions. I take this position advisedly after having studied the issues extensively for more than three decades and after having participated in governmental and academic fora during the same period. The Physiocratic doctrine isn't dead in the West! Nor are Japan's agricultural society and trade system in for sudden and major transformation. And so on.

Having pursued these subjects vigorously in the 1970s, I decided in the late 1980s to revise and update earlier research. Helpful in the process were particular events, assignments and people. For example, my association with the International Agricultural Trade Research Consortium (IATRC), a group of academic and government scholars from the United States, Canada and Europe, brought great dividends. The same can be said of the US-EC Agricultural Conference, sponsored in the United States by the US Chamber of Commerce, and of my special study of Japanese agricultural policies for the US-Japanese Commission. The former Trade Policy Research Center (TPRC) of London and its director, Hugh Corbert, offered advice along the way, as did the late Jan Tumlir, Director for Research in GATT, Geneva.

However, it is principally to my colleagues and co-workers at the University of Arizona that major credit is due. Of invaluable help was Mark Lynham's "leg work" in agricultural attaches' offices in foreign embassies in Washington, and in a variety of European locations where he gathered basic raw materials on the red meat trade. Anthony Crooks assisted on the question of barriers in the poultry trade. My special research assistant, Robert Rothenberg, was indispensable in a variety of ways, including completely reworking an earlier draft, which is hardly recognizable in the final version! Dr. Lawrence Witt wrote material on agricultural and trade policies that became the basis for Chapter 8.

Editorial assistance is the ultimate ingredient for any good manuscript. During the last two years of its gestation Thelma Leisner in London worked and reworked the manuscript and was merciless in eliminating extraneous material, and she provided helpful suggestions on content. Jenny Gray compiled the Index, Jane Wershay checked the Notes and Bibliography, and Linda Phipps did the figures; my administrative assistant and personal secretary, Joan Kingan, has been constantly on call.

Finally, there is the person who is always ready to go above and beyond the call of duty to get the typescript completed. In this case that person is Elnora Fairbank. Without her efficient help and single-minded devotion to duty the manuscript would have been delayed considerably. I owe her a special debt of gratitude.

On the personal side, during the past several years, my wife, Helen, has tolerated my anxiety over this work. She deserves that special place in that pantheon of personages which only a close companion can occupy.

Jimmye S. Hillman

Biographical Note

Jimmye S. Hillman is professor of agricultural economics, University of Arizona. For twenty-nine years he was head of the Department of Agricultural Economics and from 1986 to 1988 was director of international programs.

His major research interests are interstate trade restrictions, international barriers, and nontariff trade barriers (NTBs). These and related works have provided background material for the first realistic negotiations on NTBs in the Uruguay Round of the General Agreement on Tariffs and Trade (GATT). He served as executive director of the National Advisory Commission on Food and Fiber and, at the request of the Presidential Commission on United States-Japanese Relations, Dr. Hillman led a team and conducted an agricultural policy analysis that has promoted bilateral understanding and has assisted in reducing Japan's trade barriers. He, with the help of several colleagues, organized the International Agricultural Trade Consortium.

He is recipient of the United States Department of Agriculture's International Honor Award for 1983.

Introduction

In the summer of 1990 the British "mad cow" disease made Britain mad at France, which had rejected demands that it lift its ban on beef from across the Channel. British officials were irate when France imposed the temporary ban on British beef. The French said the beef was tainted by bovine spongiform encephalitis, the nervous disorder known as "mad cow" disease. This and the 1988–89 United States-European trade confrontation over hormones in beef are symptomatic of the tendency of countries to discover protectionist devices, other than the traditional tariff, to restrict the movement of agricultural products in international trade. These devices go by many names: for example, technical barriers, or nontariff barriers.

The failure of the major trading partners of the world to reach an accommodation at Montreal in December, 1988 on agricultural trade issues delineated in the Ministerial Declaration on the Uruguay Round of GATT, and the continued posturing of agricultural politicians in Geneva are indicative of a *no solution* outcome of world agricultural trade policies at loggerheads. Longtime observers and students of agriculture and international commerce should not be surprised. Protectionistic as well as autarkic tendencies are centuries old and are deeply imbedded in the agricultural sectors of the world. Modern nation states have not succeeded in eradicating these fundamental tendencies. In fact, this study will show that through interventionist legislation of these states and the resultant propensity of their bureaucracies to create and implant rules and regulations, agricultural protectionism has become more subtle and imbedded. Who can deny that agricultural protectionism has not decreased in real terms since World War II?

The progressive metamorphosis of protectionistic devices from the mercantilistic practices of the 17th and 18th centuries, to the protective

1

tariff during the Pax Brittanica, to the regulatory maze of the late 20th century trade is evidence of continued attempts on the part of businesses—and governments—to gain advantage by means other than economic efficiency.   The growth of such attempts was particularly noticeable after the breakdown of world trade in the 1930s. Indeed, as we shall see, the very nature of protectionistic devices changed because of the products and processes which modern industry has created, and because of the types of government intervention that grew to cope with industrial change. The growth of so-called nontariff trade barriers since the 1920s represents a new era in protectionism, and their multiplication since World War II has replaced customs duties as the most important form of agricultural trade distortion.   In fact, they now represent the major threat to both international cooperation and the future of efficient agriculture in the world.

Recent trends in the General Agreement on Tariffs and Trade (GATT) as well as positions taken by major agricultural trading nations such as the United States corroborate this state of affairs. The Declaration on the Uruguay Round of Gatt by ministers meeting at Punta del Este specifically pointed to nontariff measures and agricultural trade as subjects for negotiations; and the Declaration is replete with language which deals with regulatory matters. The United States position (see Appendix A) gave especial attention to health and sanitary measures as being an area in need of harmonization and dispute settlement.

The growth and proliferation of technical and nontariff barriers in agriculture have derived principally from the intervention of government at all levels—local, regional, national—ostensibly to facilitate the trading process, and thereby to increase economic efficiency and to improve agricultural income. The ultimate dilemma of intervention arrives when regulator comes into conflict with his clientele, or those on whose behalf he is working—consumers as well as producers. Many of the restrictions on trade in modern industrial agriculture provide prime examples of this conflict—import quotas, and variable levies being the most obvious. But there are many others that get little attention until an apparent trade violation occurs—packaging and container specifications, bilateral agreements and state trading, licensing arrangements, import calendars, mixing regulations and, of course, health and sanitary regulations. Legislative and regulatory measures relating to all of these are more subtle and subject to administrative discretion.   Such measures also are more difficult to research as to intent, and to evaluate with regard to their economic effects. One is referred again to the current (1989–90) fight between the United States and the European Community over hormones in beef which is a classic example. Is this a measure being used by the European Community to protect the European beef producers from competition?

Often the result of nontariff protection is clearer than the intent. Very few governments would admit that legislation is specifically designed to protect an inefficient producer or distributor. Neither would an enforcer of regulations—a regulator—likely admit that his job is simply that of an auxiliary or that he has been "captured" by the industry which he regulates. Nor would the regulator readily admit that he has different standards of performance, depending on the occasion, the customer or time period; and so on.

Analyzing important and underlying reasons for the growth and proliferation of agricultural protection, one discovers that overt discrimination is not expedient in the context of current trading conditions. Hence, governments have turned to domestic farm policies and programs to raise and stabilize prices and to raise incomes, all of which require intervention in the trade process. If all other techniques fail they resort to export subsidies of various nomenclature. Recent experience of the European Community with export subsidies and of the "export enhancement" device of the United States leads one to fear a worsening of this form of protection. The principal hope is that the tremendous cost of such actions will, of itself, inflict mutual corrective action and will induce remedial legislation on the part of violators.

Domestic farm programs are built into the legislative processes of the modern industrial state. The rationale for their actions varies from country to country. Price and income supports, rural employment, food self-sufficiency, protection of health and sanitation are the principal reasons given. But there is an even stronger underpinning for agricultural protection: the psyche of those who believe that farmers and agriculture, being biological and dependent on Nature, somehow are different and deserve special consideration. One has but to encounter the latent physiocratic mentality in Western Europe, the farm fundamentalism in rural United States, and the officers of Japanese rice cooperatives to quickly discover that it is somehow God's wish that the *status quo* in their agriculture be not only defended, but also preserved.

In any event, recent experiences, including the Montreal confrontation, and the Houston Summit demonstrate that the major producer countries appeared not very willing to negotiate seriously about lowering protection in their agricultural sectors. That inference seemed to continue in late 1990 as the Uruguay Round of GATT proceeded toward its appointed termination in December. Even if direct intervention measures were greatly altered, reducing the effects of various nontariff barriers will be even more difficult of achievement.

The reasoning for this is that agricultural producers and agribusinesses will not only use the more obfuscating route of nontariff restrictions to protect themselves but, along with this, they have great allies in the government bureaucracies which have been put in place over the past half century. Thus, there is mutuality in the desire of both groups for self preservation. Techniques for protection are constantly being invented by both parties and implanted in the world's centers of government.

Protection is not always an international phenomenon, witness the implications of United States sugar policy for its domestic corn producers and sugar users. Also, in one chapter in this book there is an in-depth treatment of the French-English poultry trade war of the early 1980s, a classic case study of nontariff protection by the English poultry trade. There continues to be frequent intra-EC clashes over wine, lamb, oilseeds, and even over the recent beef hormone issue.

The final chapter of this work will describe current efforts to negotiate technical and nontariff barriers as well as with the entire question of agricultural protection. As stated there, one discovers a cause and effect relationship between reducing tariffs and other forms of agricultural protection and the rise of nontariff distortions of trade. Efforts to limit nontariff barriers through the GATT and through bilateral discussions have often proved ineffectual. Analysis and measurement of these barriers have been difficult, which complicates matters still further.

Progress on reducing nontariff barriers will be exceedingly slow in agriculture unless substantial adjustments are made in agricultural policies of the major temperate-zone producers: the United States, Europe, Japan, Canada, Australia and New Zealand. If some accommodation on legislation could be found by these countries, that would, indeed, be progress: but many other countries must follow suit. In the final analysis, however, the regulator and administrative personnel in all countries—those in charge of overseeing and enforcing legislation and the rules—will have a major role in the future of the nontariff question.

1

Use of Nontariff Measures
in Agricultural Trade

As of 1990, the world had completed two of the most remarkable decades in the history of international commerce. Soon after the two United States devaluations of October 1971 and March 1973, ending the Bretton Woods era, a broad set of anomalous agricultural and food circumstances set in, accompanied by the first large oil price hike. The Tokyo Round of trade negotiations was inaugurated amidst all this. The world food crisis, the unprecedented use of export limitations and embargoes by the United States, and a greater awareness of dietary deficiencies in developing countries raised public and political consciousness of world food problems, creating a sense of optimism among many agricultural producers about future export prospects.

The second oil crisis, commencing in 1979, sent shock waves through legislative assemblies and business communities and greatly accelerated the emergence of the international debt problem that proliferated in the mid-1980s. The much altered economic environment brought inflation, severe in both industrial and developing countries, and a rapid rise in trade in both industrial and agricultural products. Though the latent ghost of inflation still stalks the halls of monetary authorities and financial markets, expanding production and slowing increases in effective demand brought a halt to trade expansion in the mid-1980s, and, for some products, actual trade declines. Nor is this all. Hundreds, perhaps thousands, of other smaller economic and political events occurred in the two decades past to make of that period a watershed, a resultant new economy consisting of some fundamentally new ingredients. (Japan's economic strength, China's willingness to trade, the role of the USSR as a more permanent grain importer, the crumbling of East-West barriers in Europe, and the change of the United States to a major debtor nation are a few examples.)

Meanwhile as the Uruguay Round began in 1986, GATT became more and more irrelevant as protectionist sentiment continued to grow. The post-World War II international economic system, which provided the framework for an unprecedented quarter century of expansion in world income and trade, however, began to crumble with the events that began in the early 1970s. International monetary arrangements were racked by successive crises: first, the breakdown of the international Monetary Fund (IMF) system in August 1971, and the second amendment to the IMF articles of agreement. Second, along with other economic aberrations, the international debt crises of the early 1980s required loan write-offs, refinancing plans and drastic adjustments in other financial arrangements. These events had an impact on many economic sectors, including agriculture, and the politico-economic role of international investment decisions provoked about as many controversies as did strictly trade questions. The nature of the "new" economic system (if it can be called a system) contains larger elements of risks and uncertainties—in both the trade and the financial arenas.

Although there have been ebbs and flows in the degree of public acceptance of moves toward liberalization and protectionism, usually related to levels of prosperity and the prices of food and raw materials, producer groups have been able to transform protectionism from traditional tariffs into more complicated, yet still effective, forms of restriction, epitomized by technical and nontariff barriers.

In effect, groups seeking protectionism were forced to invest, or to give more attention, to nontariff barriers as earlier rounds of trade negotiations eliminated remaining levels of traditional restrictions. The Uruguay Round faced this issue head on. Nontariff barriers are likely to receive additional attention in the future, for reasons that are implicit in the structure of agricultural policies being followed by most industrial nations, particularly Japan, the European Community and the United States. This structure induces producers, bureaucrats, and politicians to seek ways to increase exports for some products and reduce imports for other products, depending upon the particular circumstances of the nation states involved. That issue was also faced forthrightly at Montevideo.

Despite the apparent pessimism of the above paragraphs, it is also true that many sectors and countries have become increasingly interdependent across international lines. The economic integration of Western and Southern Europe (if not with the rest of the world), the state of near free trade between the United States and Canada, the special access for the poorest developing countries, and the integration and dependence of the "little dragons" of the Western Pacific with the world economic scene, all contrast with the complaints about setbacks in the growth rate of exports and of importers' refusals to liberalize trade restrictions more quickly. The burgeoning populations in developing countries would import more food and consumer goods were foreign exchange earnings more available; a potential for further growth in trade exists, with improved international financial relations. Dislocations in farm production reduce purchasing power for both industrial and developing countries and worsen world financial relationships. Add to this potential for trade improvement, the closer economic relationships that seem to be emerging between the Eastern Bloc and the West; opportunities exist with wise, long range economic policies. A philosophy of trying to protect existing markets leads to chaos and an inability to reach for greater potential.

National Agricultural Policies

In the current trading climate overt discrimination against imports would not be expedient. Governments therefore rely on domestic agricultural policies to stabilize prices and to raise agricultural incomes. This in turn, however, may lead to surplus output in which case governments resort to export subsidies of various nomenclature. Recent experience of export subsidies in the European Community and of "export enhancement" in the United States leads to expectations of a worsening of this form of protection. The principal hope is that the heavy cost of such actions will enforce corrective action through remedial legislation.

Most industrialized countries support their agricultural sector in some way. The rationale for their actions varies from country to country. Price and income support, rural employment, food self-sufficiency, protection of health and sanitation are the principal reasons given. But there is an even stronger underpinning for agricultural protection: the psyche of those who believe that farmers and agriculture, being biological and dependent on "nature" are different and deserve special consideration. In Western Europe, the latent physiocratic mentality can be encountered; in rural parts of the United States farm fundamentalism is much in evidence; and similar attitudes exist in the Japanese rice cooperatives. It does not take long to discover that the *status quo* in agriculture must be not only defended but also preserved.

Recent experiences at international gatherings demonstrate that the major producer countries are not very willing to negotiate seriously about lowering protection in their agricultural sectors. Even if agreement could be reached on a reduction of direct support to agriculture, nontariff measures might still be used. Agricultural producers and agribusinesses will do all they can to protect themselves through the use of nontariff restrictions.

Nontariff barriers and other types of protection are not confined to agricultural producers in the temperate zone. They have been grafted in a variety of ways on to the economies of the developing world. While developing countries have called on industrial countries to reduce restrictions and to open markets for such products as sugar, tropical products, fruits and vegetables and even livestock products, they have, in turn, been guilty of attempts to restrain trade through price stabilization and production control measures. The United Nations Conference on Trade and Development (UNCTAD) has organized specific efforts to protect and promote a number of agricultural and raw material products. Its main technique, that of commodity agreements, has been quite ineffective in achieving substantial progress. Industrial countries, while making some gestures toward assisting the developing countries with their agriculture, have not been greatly affected by this form of protection.

Definition of Terms

The language used to discuss trade protection is sometimes emotive and too often obscure as well. As was observed by one philosopher, the purpose of language is as much to conceal thought as to reveal it. Other researchers have dealt with the semantic problems involved in describing nontariff measures.[1] In this book, the term "nontariff measures" will include all those restrictions other than traditional customs duties which distort international trade, such as impediments at national borders, all types of domestic laws and regulations which discriminate against imports as well as subsidies aimed at stimulating domestic production. A "nontariff measure" is defined as any device or practice other than a tariff which directly impedes the entry of imports into a country and/or which discriminates against imports—that is, does not apply with equal force on all domestic production or distribution.

It can be argued that the intent of a government introducing a particular practice is the most important test of whether it is a nontariff measure. Intentions determine whether the measure or practice is used specifically as a tool of trade policy or whether the distortion is an ancillary effect. The Committee for Economic Development in the United States has aptly summed this up:

> National governments sometimes have pressing economic and social reasons for adopting domestic measures which may result in trade distortions. For a number of these policies the

traditional view has been that they are of strictly domestic interest and not a matter of concern for other countries . . . First of all international agreements relating to NTBs are likely to be opposed by governments on the grounds that they would limit a government's ability to deal with domestic problems. Secondly, with nontariff distortions . . . it is difficult to match and balance the concessions made by each country.[2]

Nontariff trade restrictions, therefore, usually involve some legitimate exercise of a state's authority to regulate its domestic commerce for the health, safety and well-being of its citizens. These regulations may include written or unwritten potential barriers to trade. Harald B. Malmgren, formerly United States Deputy Special Representative for Trade Negotiations in the United States, has described the scope of nontariff measures as follows:

In essence, nontariff measures are usually linked to domestic economic and social objectives: promotion of employment in low-wage unemployment areas; protection of key political constituencies; protection of priority sectors in national growth plans; preservation of government control of production and distribution of certain types of products; protection of minimum production capability of national security reasons; import substitution for balance-of-payments reasons; protection of consumers; and similar objectives. Sometimes the nontariff measures are intended by governments to discriminate against foreign trade. At other times the nontariff measures have come about unintentionally, often as an outgrowth of some domestic policy decision, which never took the trade impact into account at all.[3]

Growth of Nontariff Measures

In the last twenty-five years governments, commercial traders, economic and political analysts have come to recognize that nontariff measures present far greater obstacles to the flow of goods and services than tariffs, which have actually been reduced to very low levels. The issues involved are complex; they reflect the nature of the revolution in production technology and in the distribution of goods and services that has taken place since World War II. They also reflect the seeming inability of man and his traditional institutions to adjust to new developments and new conditions.

The GATT has drawn up an inventory of nontariff measures consisting of over 800 separate items, based on notifications submitted by governments and prepared after consultation with traders. This inventory was part of the preparatory work for the Tokyo Round of trade negotiations (1973-79).[4] Nontariff measures were the main focus of the Ministerial meeting of the GATT in November 1982 at which the ministers decided that each signatory member would compile a revised list of nontariff agricultural trade barriers and submit them to a meeting at a later date. Just before and during this meeting, however, a plan was circulated among the participating governments which called for a freeze on the level of all trade barriers; this freeze was to be followed by a gradual dismantling of the remaining tariff and nontariff trade barriers. The effort was to no avail. The meeting produced rhetoric and empty promises, tending to exacerbate rather than allay the conflict of interests over nontariff measures.

Nevertheless, concern over nontariff measures has increased for three major reasons. Prior to World War I, international trade was dominated by the movement of relatively few agricultural and related raw materials on the one hand and a few standardized items of capital equipment on the other. In the years since then, thousands of new products have been produced and the form of the old products has changed. This has created a need for new standards, new health and safety regulations and other new control mechanisms. The international transmission of technology has also led to problems of economic adjustment. A rapid transfer of technology from one country to another may result in improved efficiency and lower costs of production for an item produced in a second country. A consequent reduction in the level of protection for the product in the second country might be expected. But this does not always happen because inefficient producers in that country generally insist on protection against other suppliers (usually foreigners) so that they do not have to make difficult adjustments.

The second major reason for the increased interest in nontariff measures is that they are much more visible and their effects are felt much more strongly, now that direct protective measures such as tariffs have been greatly reduced or even removed altogether. It has been suggested that the removal or reduction of tariffs has an "incitement" effect encouraging the introduction of new, or the extension of old, nontariff measures in order to maintain previous levels of domestic protection.[5] Certainly the removal of tariffs and the reduction of their importance in international agricultural trade has focused attention on the use of nontariff measures.

A third reason for the recent increase in importance of nontariff measures is the rapid growth of public agencies since the 1920s. These agencies administer and sometimes introduce the large number of regulatory activities which exist in international commerce. The existence of these agencies has made it easier for governments to influence economic policies in indirect ways which are difficult to define or measure. There is evidence that administrative power has been used to discriminate against foreign competition in many areas—for instance, in negotiating for public purchases of goods and services, in implementing import quotas, in negotiating voluntary bilateral export controls and in administering the vast complexes of state trading.[6] Because of the complexity of nontariff measures and the difficulty of identifying some of them, they have often been ignored in international trade negotiations. There is a tendency for negotiators on trade problems to deal principally with policies related to tariffs, prices and other quantitative indicators and to leave in abeyance the more abstract and complicated subjects.

The growth in the application of nontariff measures to agriculture means that the industry will be slow to adjust to changing conditions of production and trade. In addition, nontariff measures could lead to lengthy disputes between countries. More than a decade has passed since D. Gale Johnson identified the dangers of the "disarray in world agriculture," but the portents are, if anything, more immediate now and the warning has a prophetic ring.

> It is difficult to overestimate the dangers of current trends in agricultural protectionism to the future of trade liberalization generally . . . . The 1970s may see a reversal of the trend towards trade liberalization that has persisted for almost four decades. If it does it will be to some considerable degree because no progress is made to reduce barriers to trade in agricultural products.[7]

Progress on the reduction of nontariff measures is likely to be exceedingly slow in agriculture, however, unless substantial adjustments are made in the agricultural policies of the major temperate-zone producers: the United States, the European Community, Japan, Canada, Australia and New Zealand.

Outline of the Book

The following chapters take up in more detail the problems which have so far been outlined. In Chapter 2, the origins of protection in agriculture in the United States, the United Kingdom and one or two other countries are traced. In Chapter 3 consideration is given to some of the factors which influence the use of trade controls in agriculture as well as the institutional barriers to trade and in Chapter 4 the different types of nontariff measures are discussed.

A brief outline is given in Chapter 5 of the use of nontariff measures by the United States, the United Kingdom, the European Community and Japan. Chapters 6 and 7 are case studies on (i) the nontariff measures applied to livestock and red meats and (ii) the dispute between France and the United Kingdom over imports of poultry. In the final chapter the outlook for the abolition of technical barriers to trade is assessed. The priority which might be given to certain types of action by the GATT is considered and suggestions put forward for taking steps to dismantle at least some of the more protectionist technical barriers to trade.

Notes and Sources

1. In addition to the many hundreds of shorter articles and references, a few of the major studies which grapple with the subject are: Robert D. Baldwin, *Nontariff Distortions of International Trade* (Washington, D.C.: Brookings Institution, 1970); Gerard and Victoria Curzon, *Hidden Barriers to International Trade*, Thames Essay No. 1 (London: Trade Policy Research Centre, 1970); Gerald and Victoria Curzon, *Global Assault on Nontariff Trade Barriers*, Thames Essay No. 3 (London: Trade Policy Research Centre, 1972); Geoffrey Denton and Seamus O'Cleireacain, *Subsidy Issues in International Commerce*, Thames Essay No. 5 (London: Trade Policy Research Centre, 1972); *United States International Economic Policy in an Interdependent World*, Vol. 1, Section 5, Part 2, published for the Commission on International Trade and Investment Policy (Washington, D.C.: U.S. Government Printing Office, 1971), pp. 617–90; and *Nontariff Trade Barriers*, Publication No. 665, published for the United States Tariff Commission (Washington, D.C.: U.S. Government Printing Office, 1974).

2. *Nontariff Distortions of Trade* (New York: Committee for Economic Development, 1969), p. 10.

3. Harald B. Malmgren, *Trade Wars or Trade Negotiations: Nontariff Barriers and Economic Peacemaking* (Washington: Atlantic Council of the United States, 1970), p. 13.

4. See *GATT Activities in* 1981 (Geneva: GATT Secretariat, 1982) pp. 21-22.

5. Denton and O'Cleireacain, *op. cit.*, pp. 2–4.

6. During the Tokyo Round of trade negotiations, contracting parties to the GATT agreed to liberalize numerous trade procedures, some of which related to government procurement. "The [GATT Agreement on Government Procurement] applies to government entities and agencies that each country agrees to list. Purchases of agricultural products are generally not covered."

"Government procurement was an area specifically excluded from the GATT. The Government Procurement Code does not seek merely to adjust or clarify existing GATT provisions."

"In particular, purchases by governments of supplies for government use are outside the scope of the national treatment and most-favored nation obligations of GATT. Trading for other than governmental use through state enterprises are subject, under GATT Article XVII, to the general principles of nondiscrimination." *Report on Agricultural Concessions in the Multilateral Trade Negotiations*, FAS-M-301, USDA (Washington: Government Printing Office, 1981), p. 8.

7. D. Gale Johnson, *World Agriculture in Disarray* (London: Fontana-Collins, 1973), pp. 24–27.

2

Agricultural Protection in Historical Perspective

The distortions in trade in agricultural products are one of the major problems in international trade. In the historical overview with which this chapter begins, it will be argued that the modern farm problem first appeared in Western countries about 100 years ago during a sustained decline in the terms of trade for agriculture. Since then there have been continual challenges to the commitment of national governments to protect the agriculture industry. Many of the instruments of protection which have become familiar today first made their appearance in the 1920s and 1930s. There has, nevertheless, been a marked change in the nature of agricultural protectionism. Whereas tariffs once played the predominant role, they have gradually been superseded by nontariff measures. As a consequence, governmental responsibilities in the sector have increased and whereas farmers have, on the whole, become more prosperous, they are also more dependent on government than ever before.

Origins of the Trade Problems in Agriculture

In the early nineteenth century, the United States historically had a comparative advantage in primary products and for the most part, the agricultural sector did not need protection; the tariff traditionally was the tool of manufacturing interests. The great agricultural depression which lasted from the 1870s to the 1890s gave rise to an upsurge of rural discontent and a demand for government assistance and in particular for protection for farm as well as non-farm interest under the trade laws. "Make the tariff effective for agriculture" was one of the slogans around which farmers rallied at the time.

14

As a result a detailed protective policy towards agriculture was introduced in the 1890s when a complete schedule of duties on farm products was adopted. The so-called McKinley Tariff of 1890 represented a coalition between the industrialists of the eastern United States and the grain and livestock producers of the western states; this alliance paved the way for high duties on agricultural imports for more than 40 years. In retrospect, the highly protected position of the American farmer during that era is difficult to rationalize on anything but political grounds, for most tariffs were ineffective in solving the price and income problems of the low-cost agricultural producers.

The agricultural depression of the late nineteenth century also affected Western Europe. The opening up of the American (United States and Canada) hinterland for the production of grain and of livestock, coupled with improvements in technology and transport, exposed European farmers for the first time to overseas competition. Agriculture had become an international business, in which they found themselves at a distinct competitive disadvantage. Wherever agricultural interests commanded a position of political power, however, this disadvantage was offset through the imposition of customs duties. In France the "Meline Tariff" of 1892, named after the Minister of Agriculture, imposed the same high duties on agricultural imports as on imports of manufactured products. Germany's protective policy was particularly directed at foreign grain, on which it applied high and rising duties in the tariff revisions of 1878–1902. In Italy, Belgium, and Switzerland imports of foreign grain, livestock and livestock products were taxed heavily. From 1870 on, tariffs were increased on agricultural products entering Austria-Hungary, Sweden, Spain and Portugal.

As in the United States, the agricultural protection movement in France in the late nineteenth century owed its success largely to an alliance of the agricultural and industrial sectors. In both countries manufacturers had large internal markets to serve and their concerns were primarily insular. The French manufacturers could afford to pay their workers high wages because they already enjoyed protection in the home market and depended little on export markets. The rural population was large in France and in spite of a series of "democratic revolutions," the landlord class which organized and ran the principal farm lobby, still exercised a powerful influence in the Third Republic. The absence of an organized academic opposition to protection, such as existed in Great Britain, facilitated the growth of agricultural protection in France. Since the Physiocrats, French social and economic thought has been partial to agriculture and the writings of Jules Meline provided a popular justification for its preservation.

The success of agricultural protectionism in Germany may likewise be attributed to the power of rural interests, but a coalition of industrialists and the aristocratic landowners of Prussia (Junkers) was only tenuous and intermittent, a marriage of political convenience without underlying economic rationale. The organization of great landlords in 1893, the *Bund der Landwirte*, managed to engineer the downfall of one chancellor (Caprivi) committed to liberal commercial reforms. The Bund also intimidated other opponents and made it possible to maintain price support for staple foods in spite of the fact that German manufacturing industry depended upon competitiveness in export markets. Agricultural protectionism in Germany benefited from the trenchant advocacy of professional economists such as Adolf Wagner, who drew attention to the importance, for the populous, trade-dependent Germany, of an agricultural sector large enough to guarantee food supplies and to provide employment. Another German economist, Friedrich List, provided theoretical support for the concept of tariff protection.

Britain stands out as the major exception to the wave of protectionism in Europe. When the British government abolished the Corn Laws in 1846, it committed itself to a policy of cheap food. When many of its trading partners began to erect tariff barriers and when cheap American grain brought disaster to the British landed interest, the protectionist movement made a number of attempts to establish itself on British soil without much success. In 1906, however, the Tariff Commission was set up by the government and it recommended subsidies for British agriculture and warned of the danger of dependence on imported food. Nevertheless, no major party in Britain embraced the cause of protectionism.

This indicates the way in which British economic interests, economic thought and political structure differed from those of continental Europe in the nineteenth century and in the early years of the twentieth century. Britain's economy depended on foreign trade and investment and the British manufacturing industry had enjoyed competitive superiority so that there was no need for tariff protection. Food security was not a serious concern for Britain, notwithstanding the Tariff Commission's admonitions to the contrary, because of abundant supplies from the British Empire. In addition, the economics profession was dominated by articulate advocates of free trade; their influence on public and official thought had been strong for many years.

Protection in Agriculture up to 1930

Various attempts have been made to measure the level of tariff protection given to agriculture up to the 1930s. This is not easy because of

TABLE 2.1   Average "Potential" Tariff Levels for Foodstuffs[a]
(Percent)

Country	1913	1927	1931
France	29	19	53
Germany	22	27	83
Italy	22	25	66
Belgium	26	12	24
Switzerland	15	22	42
Austria[b]	29	17	60
Sweden[c]	24	22	39
Finland	49	58	102

Source: M. Tracy, *Agriculture in Western Europe: Challenge and Response*, 1880–1980 (London: Granada, 1982) pp. 25–130.

a: These figures represent the unweighted averages of duties on 38 important   foodstuffs, expressed as a percentage of the export prices of leading European exporting countries.   Hence the results cannot be regarded as precise, and the increased incidence of specific duties reflects the decline in prices as well as the increase in duties themselves.   Nevertheless, the figures indicate the relative levels of prevailing tariffs.

b: For 1913, the figure given is for Austria-Hungary.

c: Fruit and vegetables not included.

the statistical problems involved in translating "specific" duties to *ad valorem* nominal levels of protection.   Nevertheless, a comparison of the levels of tariffs on foodstuffs in certain European countries in selected years has been made and the results are shown in Table 2.1.

As can be seen from this table, by 1931 tariffs had been raised in many countries to several times their levels before World War I.   In addition, Britain adopted a policy of protectionism in the 1930s and the United States had passed a succession of tariff acts, culminating in the most prohibitive measure of all, the Hawley-Smoot Act of 1930.   High specific duties coupled with the steady decline in agricultural prices on world markets during the 1920s resulted in *ad valorem* tariff equivalents for some commodities of 200–300 percent of their world market prices.

Although tariffs were the main instruments of agricultural protection in the late nineteenth century and the early years of this century, a few nontariff measures that have since become standard components of the modern protectionist arsenal made their appearance at this time.

For example, to enable beet sugar farmers to compete with the cane sugar growers in colonial lands, France began to give export subsidies to beet sugar farmers in 1884. Other European countries followed suit. In 1902 the dumping of surplus sugar from all quarters finally forced a number of countries to convene a conference at Brussels, where they collectively agreed to forego export subsidies on sugar.

A further example is found in the meat and livestock trade. The high tariffs on the imports of grain into Europe placed livestock farmers at a disadvantage relative to their counterparts in the United States. Transport costs were declining and refrigeration was being introduced which meant that competitive exports of live animals and meat from the United States were beginning to find their way into European markets. An effective remedy was found in sanitary embargoes. There were certainly sound biological grounds for this precaution for American government inspection services were generally unreliable, if available at all. The government of the United States, therefore, developed inspection standards and procedures for its cattle industry in order to regain access to Europe. Nevertheless, for some time after the danger of infection had been eliminated, France and other countries retained the sanitary embargo.[1]

Germany at this time excelled in the art of nontariff protection. By 1900 live animal imports were more or less prohibited as were canned and chemically-preserved meat and sausasge. Pickled or salted meat entering the country had to be shipped in pieces of prescribed weights. Fresh beef could enter only in whole carcasses, fresh pork only in half casrcasses and both often only at certain ports and on certain days of the week. Imported meat was subjected to an inspection that domestic meat was not required to pass and a high fee had to be paid for this inspection.[2]

Protection in the United States since 1930

The relationship between the governments of Western countries and their farmers in the twentieth century has been characterized by attempts to raise and to stabilize income and to develop self-sufficiency in the basic crop and livestock sectors. Governments have tried to protect the agricultural community from the wide fluctuations of world commodity markets. This led to an ever-increasing government involvement in agricultural production and trade.

Prices of agricultural products had been depressed for some years before the American stock market crash of late 1929 which set off the downward spiral of depression for the economy as a whole. The plight of American farmers became acute in this period and the government went to great lengths to support them. The average *ad valorem* tariff equivalent under the new Hawley-Smoot Act of 1930 was just short of 60 per cent, the highest in the history of the United States. Protection from imports coupled with the operations of the Federal Farm Board,[3] which was established in 1929, were meant to prop up domestic prices of agricultural products. These measures failed, however, not because of import penetration, but because of the relentless contraction of demand. In just over two years, from 1929 to 1931, the price of wheat fell from $1.04 per bushel to 39 cents and the price of cotton fell from 17 cents per pound to 6 cents. The Federal Farm Board did not have the funds to continue accumulating stocks while prices continued to slide; the Board was therefore abolished.

The Agricultural Adjustment Act of 1933 improved the price support system by combining loan-storage arrangements[4] transferred from the now defunct Federal Farm Board to the Commodity Credit Corporation; acreage allotments were introduced in an attempt to limit supply. Two foreign trade provisions, Section 22 of the Agricultural Adjustment Act and export subsidies for the disposal of surpluses, were to be used as needed to complement the basic price support program.

These four strands of agricultural policy in the United States together with direct payments for land diversion and producer marketing orders for perishables have remained the chief elements of the farm support in the United States to the present day.[5]

High tariffs, on the other hand, did not last. By the mid-1930s the United States had had enough experience with them to judge that, on the whole, they were not only ineffective for agriculture, they were probably harmful. The tariff revisions of 1922 and 1930 had not prevented American Farm prices from slipping well below *parity*.[6] The balance of supply and demand at home proved to be the principal determinant of farm prices. What high tariffs had done, however, was to prevent countries in Europe from repaying their World War I debts and they therefore adopted the alternative of retaliatory import controls on American goods. The tariff policy of the United States had therefore served indirectly to restrict all exports from the United States and especially farm exports. Consequently, in 1934 Congress gave authority to the President to negotiate reciprocal tariff reductions with other countries. As measured by the *ad valorem* equivalent, tariffs accordingly decreased from 46.7 percent in 1934 to 3.4 percent by 1990 (Table 2.2).

As an instrument of protection, tariffs were objectionable on two grounds. First, they were too broad to address effectively the problems that American farmers faced. United States producers in general were competitive relative to most of the rest of the world; in this regard,

TABLE 2.2   United States Tariff Reductions, 1934-1990
            (Percent)

Legislation	Size of authorized reduction	Authorized reduction From	Authorized reduction To	Average Tariff Level Dutiable Imports[a]	Average Tariff Level All Imports[b]
1934 Act	50	44.7	23.4	46.7	18.4
1945 Act	50	29.0	14.5	29.0	9.6
1955 Act	15	12.6	10.7	12.6	5.9
1958 Act	20	11.2	9.0	11.2	6.5
1962 Act	50[c]	12.3	6.2	12.3	7.6
1974 Act	60[d]	5.8	2.3	5.8	3.9
As of 1990	50	3.4	1.3	3.4[e]	3.3[e]

Sources: United States International Trade Commission and Foreign
Agricultural Service, Washington.
   a: Before tariff reduction
   b: Imports subject to duty plus those that are non-dutiable
   c: Final staging of tariff reductions was 1972
   d: Final staging of tariff reductions was 1987
   e: Lower average import prices and increased trade in higher duty
          categories have maintained average tariff levels.

they differed from their counterparts in much of Europe. Imports of a
number of commodities from particular suppliers did, however, pose a
threat at specific times and in certain markets. Second, tariffs were
blatant.   Open protectionism was interpreted abroad as naked
commercial aggression and the consequences in terms of retaliation
proved to be costly for the American economy. If tariffs were not the
answer to the problems of American farmers then other measures, which
would be more precise and at the same time flexible, were needed. The
answer was found in the use of nontariff measures or "invisible tariffs"
as they have been called by one author.[7]
   Regulations concerning health, safety and product quality were
already in operation before the 1930s and there is no doubt that some
had been introduced in order to confer economic advantages on

domestic interests. But in the protectionist climate of the 1930s depression, the temptation to use regulations to discriminate against imports was strong. It will suffice to mention two cases. By 1927, imports of meat from Argentina had grown considerably to the dismay of a beleaguered American cattle industry. In that year, evidence of foot-and-mouth contamination on a large scale in Argentina compelled the United States Department of Agriculture (USDA) to put much of the country under quarantine regulations. Under the sanitary guidelines followed at that time, the agency could lift the embargo from any meat-producing region of a country where the regulations applied once the agency was satisfied that local eradication measures had succeeded. Congress, however, was sympathetic to importunate American cattlemen and although there was no good scientific justification, a provision was added to the 1930 Tariff Act that required an entire country to be subject to quarantine regulations so long as disease existed in any part of the country.[8] This principle of country-wide embargoes still applies in 1990.

The restrictions on milk distribution in some of the Eastern seaboard states provide another example of covert economic protection during the 1930s. The right to ship milk and cream into New York State required a permit from the state commissioner of health. This was granted after inspection of the producer's facilities and certification by a veterinary surgeon. Starting in the 1920s, supplies of milk and cream from out-of-state producers were increasingly excluded from New York City by the simple ruse of sending no one to inspect the facilities of these producers. Then, by a state law of 1937 the inspection requirement was extended to milk imported from Canada; in no time at all Canadian farms ceased altogether to be able to supply milk to New York City.[9]

Agricultural Policy in Great Britain from 1930

For Great Britain, like most of its neighbors on the European continent, the tendency for nontariff barriers to grow in importance relative to tariffs also dates back to the 1930s. Tariff barriers were, in the 1930s, unable to stop the flow of low-priced agricultural products from the New World and all countries took additional measures to protect themselves regardless of existing treaty obligations. Even in Britain, Denmark and the Netherlands, which historically were free trade countries, governments resorted to a variety of nontariff measures to reduce the volume of imported goods.

Great Britain abandoned free trade in 1931 and adopted emergency protection measures which were replaced in 1932 by the Import Duties

Act. This Act remained the basic authority for the British tariff system for 40 years. Under the terms of the Act an *ad valorem* duty of ten percent was imposed across the board except that, certain major foodstuffs were exempted, including wheat, corn, meat, livestock and wool. Imperial preferences were established for the Dominions through the Ottawa Agreement of 1932. British agriculture, however, obtained little relief from the competitive pressure of the Dominions and colonies as well as other low-cost foreign producers. The policy of cheap food which had prevailed for nearly a century was difficult to change.

For this and other reasons, therefore, Britain had no systematic policy towards agricultural production and marketing during the 1930s. Various *ad hoc* measures were applied, commodity by commodity, to assist agricultural producers. The measures involved subsidies, import restrictions, marketing schemes and various combinations of these. Thus for wheat and sugar beet, assistance was given by a subsidy alone; for milk, by a marketing scheme and a subsidy; for bacon, potatoes and hops, a marketing scheme combined with import controls was introduced; and for eggs, voluntary restrictions were applied by exporting countries.

The marketing schemes introduced in Great Britain, like the Marketing Act of 1937 in the United States, attempted to restrict the supply of agricultural products in order to maintain prices and, so it was thought, to raise the income of farmers. But Britain, like other countries, found that supporting prices was a hazardous undertaking without some form of overall supply regulation. Hence, legislation to regulate imports was introduced so that the prices set by the marketing schemes would not be undercut. Marketing arrangements adopted by milk, bacon, hops and potato producers allowed the government to regulate quantities sold by registered producers and quantities purchased from abroad. Supply control afforded protection while limiting the cost to British consumers.

Britain entered into a variety of bilateral agreements with several countries during the 1930s, all of which took into consideration Britain's pledges and concessions to the countries of the Dominions and colonies. These agreements involved guaranteed maximum duties on products from Denmark and Argentina and duty-free or duty-reducing concessions on a wide range of products from the United States. Specific clauses in some of these agreements, however, gave the British government the right to impose quotas if imports began to interfere with domestic marketing schemes.

The shift from free trade toward protection during this chaotic period and a further shift toward self-sufficiency after World War II did not take place without controversy. The arguments used in the two periods were, however, very different. In the 1930s, the arguments were based on the issues of "fairness" and of the employment of domestic labor. After World War II, the case for protecting domestic agriculture rested on the dual considerations of self-sufficiency and balance-of-payments considerations. One argument was that comparative advantage had shifted against British industry and that the country would be unable to finance large food imports. Coupled with this was the fear that food prices would rise relatively faster than prices of manufactures; it was averred that this would aggravate the balance-of-payments problems. The success of these arguments in shifting agricultural policy from free trade to market intervention and modified protection is demonstrated by the fact that by 1985 Britain produced 60 percent of its food requirements (80 percent of products capable of being produced indigenously) compared with about 30 percent before World War II.[10]

Protection in Europe after 1930

It has been shown that policies in the United States and Great Britain underwent a basic change after the breakdown of world trade in the 1930s. Countries diverted their efforts from protective tariffs to direct government intervention in both production and marketing. In Continental Europe, the same tendencies prevailed. The traditional concern to maintain farm incomes was reinforced by the need to guarantee food supplies and to save foreign exchange. Old issues such as the social value of agrarian life and the structure of the community once again featured in the argument for agricultural protection.

In the depressed conditions of the 1930s, many devices to restrict imports and to support domestic agriculture were introduced in the expectation that they would be temporary and could be relaxed or removed when economic conditions returned to "normal". But normal times never materialized. The restrictive and autarkic measures taken by some European countries, moreover, led to retaliatory and defensive strategies at the international level.

France offers a clear example of a policy which progressed from tariffs to nontariff measures and then to direct measures of intervention. In the early 1930s France had the strongest currency in Europe and this was disadvantageous for French farmers. Other countries dumped their agricultural surpluses on the French market at bargain prices in terms of the "Franc Poincare," so called after Raymond Poincare, premier of France in the 1920s, whose financial policies balanced the budget and stabilized the value of the franc.   Duties were nugatory but the use of import quotas quickly spread to virtually all major agricultural products. The French government began to charge fees for import licenses both to facilitate the rationing of this valuable right and also to appropriate some of the "economic rents" that license holders were earning. The fees were adjusted according to the prevailing price differential at the frontier and in this way they acted much like the modern variable levy.

Import quotas themselves were replaced by broad state controls on wheat and wine when a tendency to over-production caused domestic prices to sag. After 1933, the French government assumed a monopoly of trade in these commodities; prices were fixed and supplies were controlled. The intervention was justified on the grounds of the long-standing importance of agriculture to the economy of France as well as the need to maintain social balance in the countryside.

The experiences of the 1930s left a permanent mark on French agrarian policy. There were those who thought that French agriculture would have ceased to exist had the government not intervened with its strongly protective measures between the wars.[11] Agricultural and commercial policies since World War II have concentrated on the need to modernize French agriculture and to raise productivity in the industry. The ultimate aim is to increase gross output and, in the process, to improve the competitive position of French agricultural producers in world markets and hence to increase agricultural exports. Faced with limited growth in the domestic demand for foodstuffs, France looked increasingly towards preferential agreements abroad, particularly with Germany and Britain, in order to dispose of its surplus production. Thus, it is understandable why sponsors of the French agricultural policy during these years also became some of the principal proponents of the common agricultural policy of the European Community.

Agricultural and commercial policies in Germany and Italy during the 1930s were subsumed under the pervasive economic management of the National Socialist and Fascist governments. In Germany, regulations controlling prices and volumes of imports were enforced on all important commodities beginning in 1933; state boards could buy and sell on the domestic market as well as operate buffer stocks. From 1933 until World War II, imports of agricultural products were severely limited.

After World War II, agricultural and trade policies in the Federal Republic of Germany were dominated by attempts to deal with problems of the structure and fragmentation of farms. Strict control through import boards helped to maintain domestic prices at levels higher than those prevailing in world markets. In spite of government assistance toward structural reform, however, German agriculture still remains highly vulnerable to foreign competition.

Italy also uses various devices to stimulate and protect its agriculture. A full state wheat monopoly was established in the mid-1930s;

growers were required to deliver their wheat to the wheat monopoly and they received a fixed price. Imports were controlled directly by organizations responsible to the Ministry of Agriculture. In the mid-1950s, soft wheat and rice production outstripped domestic demand and export subsidies were introduced in order to reduce stocks.

France, Germany, Italy and the United Kingdom are now, of course, subject to the common agricultural policy of the European Community.

Levels of Protection after World War II

As Western Europe recovered from World War II, substantial increases in agricultural production were achieved, but the trade situation became less satisfactory. Although many trade restrictions were removed in the industrial sector and although some progress was made in removing quantitative barriers to agricultural trade among West European countries, a large portion of their agricultural production and trade remained subject to state intervention and control. Between the end of World War II and the early 1960s there was, in general, a marked increase in agricultural protection, as shown in Table 2.3.

The degree of protection is measured by the percentage deviation of the domestic producer price (representing the combined effect of all protective devices) from the world price. This measurement is based on the report by a panel of experts in the GATT known as the Haberler Report.[12]

During the periods shown in the table the degree of protection in France and West Germany increased for nearly all commodities, while in Denmark, Italy and the Netherlands, increases were confined mainly to grain and milk. In the United Kingdom, apart from milk, there was not much change in the overall degree of protection for increases in one agricultural sector tended to be offset by decreases in another.

Production in Western Europe in the post-war period increased rapidly; this was the result partly of technological advances in agriculture, both in the form of inputs such as higher yielding seed-corn and of pest and disease control, and partly because of improvement in equipment and harvesting methods. Demand within Western Europe for agricultural output grew at a slower rate. In these circumstances, prices should have fallen. Governments in Western Europe, however, ever-anxious to maintain the income of farmers, used a variety of price support measures which had the effect of stimulating output still further and creating even greater imbalance between demand and supply on domestic markets. With little or no control over domestic

TABLE 2.3 Changes in Degree of Protection, 1951-61

Country and period	Wheat	Barley	Beef	Pork	Eggs	Milk
Denmark:						
1950/1-1952/3	−9	9	3^a	3^a	-4^a	-3^a
1956/7-1958/9	2	8	-4^a	-4^a	-3^a	9^a
1959/61	9	7	-3^a	-3^a	-4^a	35^a
France:						
1950/1-1952/3	7^a	—	—	28	—	18
1956/7-1958/9	20^a	5^a	-10^a	—	4	34^a
1959/61	17^a	21^a	29^a	3	19	46^a
Italy:						
1950/1-1952/3	13^b	−11	—	38	45	46
1956/7-1958/9	27^b	26	40	23	32	55
1959/61	37	28	41	23	39	62
Netherlands:						
1950/1-1952/3	−28	6	-8^a	8^a	-9^a	-6^a
1956/7-1958/9	12	16	-9^a	-4^a	1^a	32^a
1959-61	21	26	-12^a	-7^a	0^a	45^a
United Kingdom:						
1950/1-1952/3	16	9	40	36	38	36
1956/7-1958/9	9	24	42	24	34	61
1959/61	3	24	36	17	36	62
Federal Republic of						
Germany:						
1950/1-1952/3	7	10	34	24	19	18
1956/7-1958/9	29	36	45	23	32	36
1959/61	31	38	41	24	38	39

Source: Measures of the Degree and Cost of Economic Protection of Agriculture in Selected Countries. Technical Bulletin No. 1284, USDA (Washington: Government Printing Office, 1967) Table 17, p. 21.

 a: Export values
 b: Soft wheat

supply both import restrictions and export subsidies were used in order to preserve the system of price support.

In 1962 the common agricultural policy came into operation and the European Community fell heir to the intractable farm problems and entrenched protectionist remedies of its member states. In view of its inheritance, the consequences of the common agricultural policy should come as no surprise. *Plus ça change, plus c'est la même chose.* The European Community has transformed itself from a net importer to a net exporter of numerous agricultural commodities. A variety of measures have been used to deal with the marketing problems thus created, but these fall short of a consistent policy to deal with what amounts to a world-wide structural maladjustment in agriculture. If the price support systems and other interventionist techniques that were adopted after the World War II are no closer to removal now than when they were introduced, it is because, like the measures that preceded them, they have sustained and compounded the problems which they were designed to redress. The uneconomic producers have been maintained by the system but they need assistance to survive. What was observed of the early record of the quote in one French study in the late 1930s could serve as a commentary on the general history of protection.

> Protection prevents adaption, and lack of adaptation calls for protection. Consequently, unless the steps are retraced—and this appears less and less likely because it becomes increasingly painful—protective measures follow one another. This explains why, some seven years after the first quotas were introduced, we now find the quota system firmly and widely embedded in the French economy.[13]

The merits of a policy of free trade will not be argued in this context; but it is obvious that protectionism cannot succeed in the Community because it insulates the farm sector from the economic forces which induce progressive change and it diverts attention from the need for more constructive price and income policies.

Notes and Sources

1. Percy Bidwell, *The Invisible Tariff* (New York: Council on Foreign Relations, 1939), p. 170.

2. Michael Tracy, *Agriculture in Western Europe: Challenge and Response*, 1880–1980 (London: Granada, 1982), p. 97.

3. The Federal Farm Board was set up under the Agricultural Marketing Act of 1929 in an attempt to stem the tide of depression in agriculture. This was to be achieved by making loans to cooperatives and others. The primary aim of the Board was to control surpluses and at the same time to increase agricultural incomes.

4. Loans were made available to cooperatives and others, primarily to control surpluses but also to increase farm income.

5. Direct payments were made to farmers when they held land out of production and diverted it to growing crops not in surplus. Orderly marketing schemes ("marketing orders") were introduced when two-thirds of the producers of a certain crop voted for such a scheme (a system to give said producers an element of monopoly power).

6. Parity was defined as prices which would restore the purchasing power of agricultural commodities to the level which had obtained on average in the previous period, namely from August 1909 to July 1914. The concept of parity was basic to most government programs for more than 40 years. It is no longer the centerpiece of agricultural policy in the United States.

7. Bidwell, *op. cit.*, p. 1.

8. *Ibid*, pp 17-18.

9. *Ibid*, pp. 181-82.

10. *The Times* (London), July 11, 1986, p. 4.

11. See M. Auge-Laribe, *La Politique agricole de la France de 1880-1940* (Paris: Presses Universitaires de France, 1950).

12. Gottfried Haberler *et al.*, *Trends in International Trade* (Geneva: GATT Secretariat, 1958), pp. 81-84.

13. Quoted from the study by O. Long, *Le Contingentement en France* (1938), in Tracy, *op. cit.*, p. 184.

3

Institutional Barriers
to Trade

There are times when conditions are more conducive to trade controls than others.   In Chapter 2 it was shown that there have been periods when agitation for protection was more intense and the barriers to trade were increased.   Certain factors tended to influence the frequency of these periods and the intensity of the cries for protection. The first, and perhaps the most important, of these is the occurrence of economic depression—not localized or industry-specific setbacks, but general cyclical depression.   During such periods the volume of business decreases,   markets   shrink,   incomes   are   drastically   reduced   and possibilities of alternative employment for all economic inputs are limited.   At the same time, government revenue falls and government activity is reduced, causing further problems.

As discussed in Chapter 2, the dramatic increase in levels of protection   during   the   1920s   and   1930s   was   a   consequence   of   the economic disorder of Europe after World War I, followed by the depression.   Attempts were made to restrict trade in the various European markets as business conditions slackened and competition increased. During the 1930s, the drive toward nationalism and self-sufficiency all over the world was accelerated by perversities in national and world markets.   Even within the United States, where consitutional law prohibits the restriction of trade between the states, the economic conditions of that period were such that attempts were made to introduce a variety of barriers to interstate trade.[1]

At the same time, innovations in production, transportation and marketing techniques, such as the growth of mass distribution arrangements which circumvent older marketing channels were emerging.   Direct selling, the evasion of centralized markets and chain store merchandising all bring new forms of competition. In the post-World War II period improvements in the design and efficiency of vehicles and the expansion of highways have led to a rapid growth in road transport and, in many instances, road transport has replaced other forms of transportation. Specific improvements in commodity marketing techniques—mechanical refrigeration, improved methods of preserving food, standardization of packaging, revolutions in communication procedures and the streamlining of services—have all extended the area of competition and have allowed more inter-market penetration than ever before. This has led to the deterioration of local markets in many areas and, as a result, efforts have been made to protect and preserve these markets for national or local producers and distributors. Yet these same changes in production and distribution have caused increased compatibility in international trade which has facilitated its expansion.

On the other hand, fiscal concerns may also lead to increases in trade barriers. With the increasing complexity of national and local government procedures, new sources of revenue are needed to meet increased demands for funds. The imposition of a value-added tax in certain West European countries may be cited as an example. New forms of licensing and new taxes also seem to be correlated with economic crises and recessions; they may reinforce or be superimposed upon other measures which arise in periods of commercial stress. The net result is a burden on international commerce and noticeable instability in business activity.

Legislation as a Trade Barrier

Rapid advances in production and marketing often require action by legislative and regulatory bodies. Failure or delay in enforcing progressive legislation, or in revising or repealing outmoded statutes, can cause inefficiencies in marketing. For instance, marketing efficiency can be reduced as the result of a lack of uniformity in laws on sanitation, in regulations covering grades, standards, and labeling, and in requirements for specifications relating to size, weight and so on. A change in consumer demand may call for a different kind of container or a different quality item not currently allowed by the regulations. Delay on the part of legislative and administrative bodies may mean that the demand from consumers cannot be met. The intent, frequency and strength of nontariff interferences which arise in this way need not be related to economic conditions, nor deliberate attempts to protect local business from foreign competition. They may simply result from hesitation or failure to change legislation or regulations in response to changes in social and economic circumstances.

Outright retaliation against foreign imports may be directed against real or anticipated legislation in other countries, as legislative bodies try to counteract or preempt the ill effects of restrictions imposed by others. But there are probably additional forces which motivate retaliatory legislation. In times of economic crisis, an increase in retaliatory measures, as well as purely protectionist ones, can be expected, as was the case during the global recession of the early 1980s.

Administrative Protection

Concurrent with the decline in the use of tariffs and other direct forms of protection between the 1930s and the 1990s has been a dramatic increase in indirect or administrative protection, most of which has been connected with laws and regulations on nontariff issues. Administrative devices used to restrict trade include the imposition of fees and the issuing of licenses at the discretion of government officials together with the use of health and safety regulations. In all OECD countries in the twentieth century there has been a vast extension of government regulation of domestic commerce and as a result trade regulation has increased. The passing of power from legislative to administrative agencies of government has afforded the latter increasing opportunities for policymaking and interpretation. This shift of responsibilities permits decisions to be made more quickly and in more detail.

During the struggle over interstate barriers in the United States in the 1930s, it was observed that:

> In the national and municipal areas administrative action resulting from the grant of wide discretion to the enforcement officers and action extending beyond statutory authorization has perhaps been more often used for protecting local products and entrepreneurs from out-of-state competition than has purposeful legislation.[2]

There is little doubt that administrators use the flexibility of the law to discriminate against foreign competition. But it is difficult to ascertain the scope and authority of those decisions where the consequences are protectionist. There is no uniformity among countries as to the ways in which administrative directives are issued. Elected or appointed officials often have the power to formulate rules which make interpretation of legislation easier. These rules will carry the same authority as statutes. In many cases new administrations routinely issue their own revisions or supplements to the regulations. This in itself creates problems in the administrative procedure, if only because of difficulties in obtaining the latest regulation.

In the case of quantitative restrictions, such as quotas, licensing and exchange controls, the discretionary component of administration is generally small, so discriminatory or arbitrary practices are readily identifiable. In other regulatory activities abuses are more difficult to discover. To quote an unnamed official: "Honestly, it depends on the price of _ _ _ as to how rigorously I apply this particular regulation." The point is that the administration of the regulations, rather than the regulations themselves, will determine the extent of protection conferred.[3] Estimates of the effectiveness of trade barriers ought to be made by a study of actual administrative practice rather than by analysis of the powers granted by legislation. Chapters 5 and 6 present two case studies but it may be useful to cite one case here for illustration.

In the United States agriculture is regulated at three levels, by the federal government, by state governments and by producers themselves through the use of marketing orders. Under the authority granted to them by the Agricultural Marketing Agreement Act of 1937, fruit and vegetable growers currently operate some 65 federal marketing orders, setting their own standards for grade, size, maturity and packaging. They also limit the flow of the commodities to market, subject to the approval of the United States Department of Agriculture. By an amendment to the Act of 1954, Section 8(e), imports must meet the same requirements as those imposed by the producer committees on domestic suppliers. Section 8(e) appears to preclude discrimination against imports, yet at the same time it extends the powers of self-regulation to the trade.

When the Florida Tomato Committee (FTC) drew up size restrictions for the 1969 marketing season and submitted them to the Department of Agriculture for approval, the discriminatory potential of the marketing order system gained widespread publicity. The FTC proposed a more restrictive (that is, greater) minimum size for vine-ripe tomatoes than for mature green tomatoes. Their chief competitors in the American market, the growers of northern Mexico, harvested only vine-ripe tomatoes and if output of the previous year had had to meet the FTC's new standards, 40 percent would have been excluded from the American market. The FTC could claim that Florida's vine-ripe tomatoes were, on average, larger than its mature green tomatoes so that size restrictions of this kind were not unfair. But this comparison was hardly the appropriate one: Florida producers do not, by and large, sell vine-ripe tomatoes—they pick them while they are still green. The Department of Agriculture gave summary approval to the scheme

—without preliminary notice, public hearings, or the customary waiting period between publication in the *Federal Register* and implementation.

There followed a battle, which did not subside for years, to block the action in court and there was an outcry in the press. At the time it was observed: "That the USDA should go along with [the FTC] plan is a sharp reminder of the extent to which government agencies often end up serving the special interest groups that they are supposed to be regulating."[4]

However high-minded and able the administrators who conduct day-to-day policy may be, there is a danger that because they are not subject to direct public criticism and public accountability, they fall prey to the influence of domestic special interest groups. It is not that executive agencies are more susceptible than legislatures, but that this susceptibility is not limited to the elected representatives of special interests; in the realm of bureaucratic politics it escapes the publicity and constitutional checks that control it in legislative politics. As a result, in the major OECD governments agricultural interests have become entrenched. The introduction of marketing schemes, export policies and regulatory activities can often be directly attributed to pressure from organized groups of producers.

There are numerous examples of the concentration of policymaking power among administrators increasing the effectiveness of organized pressure from directly affected groups, or their lobbyists. Probably the most effective lobbying group of all is the National Central Union of Agricultural Cooperatives (ZENCHU), which is a branch of the Japanese National Farmers Cooperative.[5]

Administrative Powers in the United States

It would be wrong to infer that the executive agencies of government have become the custodians of special interests and the prime architects of protectionism. Administrative protectionism is the work of both bureaucrats and legislators. In the United States, at least, the executive branch has often been the more liberal of the two and it is the legislature that has introduced restrictive regulations despite the objection of administrators who must enforce them. This struggle, by now quite familiar, has a long history.

The President of the United States has gradually acquired responsibilities over determination of tariffs since the 1920s. There has been nothing decidedly protectionist in the way these responsibilities were exercised. In the 1922 Tariff Act Congress gave the President the power, upon investigation and recommendation by the Tariff Commis-

34   *Institutional Barriers to Trade*

sion, to adjust duties by as much as 50 percent to offset any difference between foreign and domestic costs of production. This cumbersome form of the variable levy might have been used to eliminate foreign comparative advantage over a wide range of goods. But the authority was rarely invoked. From 1922 to 1928 duties were adjusted on only 38 of 2,800 dutiable items. Congress carried over the "flexible provision" into the 1930 Tariff Act; in the next eight years only 57 adjustments were made, and more than half of them were reductions.[6] It was in these depression years that Congress gave the President power to determine tariff rates through trade agreements. That power has been renewed before each round of multilateral trade negotiations in the GATT. Franklin Roosevelt and his successors used their authority to lower tariffs to their present nominal levels.

Protectionist sentiment in Congress was strong during the depression of the 1930s and although that body deferred to executive discretion on the tariff, they found other areas of trade policy where they could limit it. Before 1930, the Department of Agriculture administered a strict embargo on cattle and red meat imports from any country or zone within a country infested with rinderpest or *aftosa*. It was the legislature which in the Tariff Act of 1930 overrode the advice of the Secretary of Agriculture and required that selective embargoes should be made countrywide.

The "Buy-American" movement that swept the states and the legislature in the early years of the 1930s offers another case of legislators rewriting administrative responsibilities in order to safeguard domestic farm interests. The federal government Procurement Act of 1933 allowed for a considerable measure of administrative discretion over the source of supplies acquired for government purposes. Domestic firms were to enjoy a great advantage, but exceptions to the policy of domestic procurement were permitted on broadly defined grounds of availability, reasonable cost and quality. When the Navy was about to accept a foreign tender for canned beef provisions, Congressmen made it clear that, in cases where American farmers were concerned, they had not intended to grant the government any discretion at all. Thus, although beef of superior quality could have been procured from Argentina at less than half the cost of the best American bid, and the President and State Department favored the deal, a coalition of the American National Livestock Association, governors, senators and representatives of a number of western states moved Congress to attach an amendment to the pending Navy Supply bill that categorically rules out all foreign supplies.[7]

Political Boundaries as Barriers to Trade

Even in the absence of restrictions which arise either because of discriminatory design or because of differences in laws and regulations, the mere existence of political boundaries can, in itself, produce nontariff barriers to trade.

Before business can be carried on in another country, many formalities (paperwork, investigations and legal procedures) must be attended to, all of which involve time and expense. Extra marketing costs produce an upward lift to the transfer-cost gradient, thus distorting the pattern of market and supply areas. Hence market areas that extend across political boundaries are reduced in size and often eliminated. It follows that the higher the costs of crossing a border more nearly will the boundaries of economic and political areas coincide with each other. If trade is stopped, the political boundary is, in all senses, an economic boundary.

The conclusion to be drawn from the above is that a most important effect of national laws and regulations lies in their influence on the location of production and processing of commodities. Figure 3.1 shows the effects of an international boundary plus the special "roundabout effect." This "roundabout effect" is best explained using the example of traffic through a border station. Since large overhead costs are involved in their upkeep, these stations, like bridges across a river, are provided only at intervals, their actual locations often being influenced by natural factors. Since all legal traffic must pass through one of these stations, the routes used for international commodity shipments are often made longer, hence slower and more expensive, apart from any extra time and expense entailed in the actual crossing of the boundary.

In Figure 3.1, producers of a particular commodity are located around A and a boundary runs along the straight line HK. With market costs directly proportional to transport costs, producers can deliver their product domestically at the same total unit cost, X unit of account per unit, anywhere on the circular arcs BC and FG. If no extra costs or distance were involved in the international transfer, the product could be laid down at the same total unit cost anywhere on the arc CSF. On the assumption that crossing the border does entail some extra costs, for example, an inspection fee or an indirect tax, but that crossing can take place at any desired point, producers at A can sell only as far as the arc DRE in the other country. The dark area represents the loss of distribution range. If regulations force interstate traffic to pass through a specific port of entry, say P, the limit of the unit of account delivery is reduced to GFMRLCB. Thus it is evident that there is a further loss in distribution range because of the "round-

36

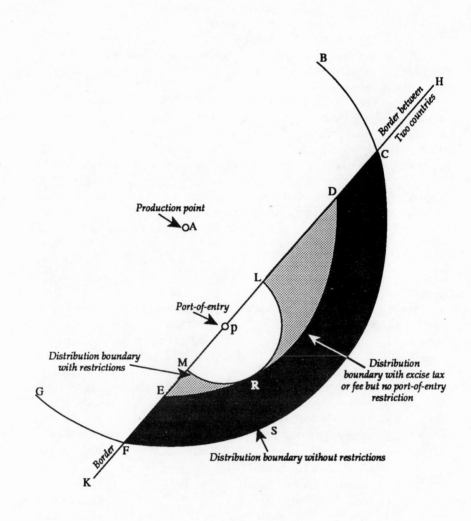

FIGURE 3.1. Effects of a Trade Barrier and Port-of-Entry Restrictions on Distribution. *Source:* Reproduced from Jimmye S. Hillman, *Nontariff Agricultural Trade Barriers* (University of Nebraska Press, 1978), p. 24.

about effect"—the funneling of traffic through P. The area lost is shown by the striped area in Figure 3.1.

The effect of a boundary is not uniform for all traded items. In the case of agricultural shipments, whether the product is processed or unprocessed, the distance of the area of production from the boundary and the susceptibility of the product to insect infestation or disease will affect international marketing and the location of production. Other factors affecting these include the uniformity of grades and standards, the relationship between bulk land value and the method of transportation used.

A return to measures regulating international trade on as large a scale as in the early 1930s would affect the location of areas of production and processing plants. Areas devoted to production near borders would suffer the economic consequences of smaller scale production, higher average distribution costs, or both. The optimum centers for the collection of inputs may be near national borders and processing plants located at these points may suffer accordingly. In order to secure a free and steady flow of materials, the industry might have to move away from the border. On the other side of the border, however, protected trade areas and production centers may be created and certain artificially created "advantages" would arise for industries which located themselves near the border in the "protected" country.

State Trading as a Nontariff Measure

The large grain purchases by the Soviet Union in 1972, the partial cancellation of American grain sales to the Soviet Union in 1974 and 1979, and the increasing importance of the People's Republic of China as an agricultural importer and exporter have focussed attention on appropriate methods of dealing with exports to and imports from centrally-planned economies. Problems involved in exporting agricultural products to centrally-planned economies should not be neglected if there is a possibility that those countries will continue to purchase large quantities of grain from the market economies.

Centrally-planned economies such as the Soviet Union and China have a considerable degree of monopsony power in agricultural markets. This power presents great problems to major grain exporters such as the United States, with its private traders, and even to Australia and Canada with their State marketing agencies. Fluctuations in the harvests in the Soviet Union have a marked effect on world demand for grain and this, in turn, has a destabilizing effect on world grain prices. One way to offset this destabilizing effect would be to control the supply of exports from all countries but this would have elements of a state-controlled oligopoly-oligopsony situation which could make the world market even more unstable.

Problems created by state trading and related monopoly-monopsony conditions of demand and supply in world trade of agricultural products can be partially resolved through better information, especially about total world supplies. Informed decisions would be preferable to the confrontation of buyers and sellers acting in ignorance. The alternative of shared information on matters relevant to potential imports and exports is discussed in a United States Department of Agriculture publication;[8] it was further explored in the World Food Conference held in Rome in November 1974.

A large part of the advantage held by monopolists and monopsonists in international trade lies in their having a monopoly of information and using it skillfully. It is this aspect which is equivalent to a nontariff measure because of the distortions caused on world markets.

The next chapter will explore in more detail the types of nontariff measures used in agricultural trade.

Notes and Sources

1. J.S. Hillman, *Economic Aspects of Interstate Agricultural Trade Barriers in the Western Region*, Ph.D. dissertation, University of California, Berkeley, 1954; Hillman and J.D. Rowell, *A Summary of Laws Relating to the Interstate Movement of Agricultural Products in the Eleven Western States*, Report No. 109 (Tucson: University of Arizona Agricultural Experiment Station, 1952).

2. F. Bane, "Administrative Marketing Barriers," *Law and Contemporary Problems*, Vol. 8, Spring, 1941, pp. 376-90.

3. For some comments on particular regulations and the implications on their administration, See Gerald and Victoria Curzon, *Hidden Barriers to International Trade*, Thames Essay No. 1 (London: Trade Policy Research Centre, 1970), pp. 26-33.

4. Quoted in Maury Bredahl *et al.*, *Technical Change, Protectionism, and Market Structure: The Case of International Trade in Fresh Winter Vegetables*, Technical Bulletin 249 (Tucson: University of Arizona Agricultural Experiment Station, August 1983), p. 17.

5. John W. Longworth, *Beef in Japan* (St. Lucia, Queensland, Australia: University of Queensland Press, 1982) pp. 58-9, 68.

6. Percy Bidwell, *The Invisible Tariff* (New York: Council on Foreign Relations, 1939), pp. 120-23.

7. *Ibid.*, pp. 255-62.

8. *Prospects for Agricultural Trade with the USSR*, Foreign Series Report No. 356, ERS, USDA (Washington: Government Printing Office, 1973), especially the article by D. Gale Johnson, pp. 43-50.

4

Types of Nontariff Measures
in Agricultural Trade

The overall objective of the General Agreement on Tariffs and Trade in 1947 was to restore orderly conditions in international commerce following the trade disorders of the 1930s and the subsequent hostilities. Of the more explicit purposes, the main one was to reduce tariffs, but another was to prevent governments from whittling away the benefits of tariff reductions through the use of nontariff methods of protection. Since the end of World War II, the GATT has provided the framework of principles and rules that have regulated the trade of the free enterprise world. Under its auspices, and despite some setbacks, a measure of trade liberalization has been achieved through successive multilateral negotiating rounds.

The special position accorded to agricultural products in the GATT negotiations, however, has meant that these successive rounds of multilateral trade negotiations have had limited impact on the long-run upward trend of protection of agriculture in the industrialized countries. Most governments have had difficulty in coming to grips with the nontariff nature of agricultural support and protection. In this respect, discussion has been hampered, *inter alia*, by lack of precise information on the subject. Attention is devoted in this chapter to the efforts made by the GATT and others, to describe more accurately the nontariff problem as it relates to agricultural trade.

Taxonomy of Nontariff Measures

Various attempts have been made to identify and classify government measures which affect agricultural production and trade. It has been observed by the Committee for Economic Development in the United States that:

> Determining which (nontariff barriers) can realistically be nego-
> tiated is a complex issue. In practice how can one distinguish
> less justifiable barriers to trade from a country's natural right
> to generate revenue through taxes that do not discriminate be-
> tween domestic and foreign sources? Or to subsidize some in-
> dustries to correct for other distortions or to impose legitimate
> product standards?[1]

One obstacle in the way of negotiations on nontariff measures is that legislative or administrative intent is not easily identified. Another problem is that, in the opinion of many investigators, the advantage that nontariff measures confer on domestic producers or the disadvantage they impose on foreign competitors "may be virtually impossible to quantify."[2] Moreover there is, as yet, no satisfactory way of evaluating the effects of nontariff measures. The attempts by international negotiators to control their use has therefore been unsystematic, risky, and ultimately ineffectual.

A basic inventory of nontariff measures was compiled by the GATT Secretariat after the Kennedy Round of multilateral trade negotiations which was concluded in 1967. Some identification and classification has also been carried out by other bodies, including the United States Department of Agriculture (USDA). The Foreign Agricultural Service (FAS) of USDA, for example, published a series of circulars in 1972-73 on its trade with selected countries in the Organization for Economic Cooperation and Development (OECD).[3] These circulars set out the types of nontariff measures used in these countries.

The United Nations Conference on Trade and Development (UNCTAD) and other writers have classified nontariff measures into three types. Type I are those where the specific intent is to restrict imports and to stimulate exports in a manner that will inevitably cause trade distortion. Type II are those measures the primary intent of which is to deal with economic, social and political problems but which are occasionally used to restrict imports or stimulate exports. Type III are measures or policies that are not intended as instruments of trade protection but which nevertheless inadvertently cause trade distortion.[4]

The nontariff measures which are most frequently used to control agricultural imports are: (a) quotas and related restrictions; (b) variable import levies; and (c) health and sanitary regulations. Each of these will be considered briefly.

Quotas and Related Trade Restrictions

Quantitative restrictions are the most important measures available to governments for the control of trade in agricultural products and for discrimination between markets. They are particularly important in developing countries where governments use them to influence the allocation of resources. By contrast with tariffs and related measures, prior import surcharges, multiple exchange rates and prior deposits on imports, quantitative restrictions break the link between domestic and world prices. Quantitative import restrictions restrict the volume of imports into a country not, like tariffs, by artificially raising the cost of importing, but by placing direct limits on the quantity (or value) of imports that may enter the domestic market, irrespective of prices.

Although quotas may be global, they are often specifically allocated among exporting countries. The point of application may vary, but usually they are applied directly to importers or to exporters in the country of origin. Some apply to processors through, for example, restrictions on the maximum amount of imported grain that may be used in domestic flour.

Restrictions are usually expressed in terms of the maximum quantities of specified commodities that can be imported during a specified time period. Sometimes, however, the specification is in terms of value. Trade agreements, or arrangements that include quota provisions, usually state the size of the quota allotted to particular countries in terms of a percentage of total imports from the countries in question.

Import quotas can be used to discriminate between trading partners, with different kinds of quotas allowing different degrees of discrimination. Guaranteed quotas or purchase agreements, wherein the importing country agrees to buy a stated minimum quantity during a specified period, are much more discriminatory in practice than "permissive" marketing grants where the exporters in the countries to which grants have been issued are not restricted in their trading practices.

In the early 1970s there was an increase in the number of cases where importing countries persuaded exporting countries to enforce and police quantitative restrictions. These voluntary export restraints (VERs), as they are commonly called, aim to shift the embarrassing task of allocating quotas from the importing country to particular exporters. In some cases, the executive side of governments in charge of such matters might be reluctant to seek and use legislative authority, or one might be reluctant to apply quotas among importers. There may also be certain administrative and economic advantages in having quotas applied in the exporting countries. For example, when the export trade is more concentrated than the import trade, quotas can be more effectively policed and shipments can be better adjusted to variations in market conditions in the importing country.   The administration of such voluntary agreements raises the question of how the margin resulting from the difference between the prices in the importing and exporting countries should be distributed.   Exploration of this question, however, is beyond the scope of this study.

The import quota, which was introduced along with other nontariff measures in the early 1930s, threw a different light on tariffs and other political instruments as protective measures.   If supply and demand for a given commodity are fairly elastic, the economic effects of the imposition of either tariff or an import quota are not very different.   If the import quota is set at the volume of imports that would result from the imposition of a given tariff, the protective effect will be the same in either case at the initial equilibrium, as will be the consumption effects. (This argument is set out more fully in Appendix B.)

The use of the import quota to control agricultural imports is widespread.   Governments feel bound to guarantee a "fair price" to domestic producers.   In many countries, policies of raising prices for agricultural products and supporting farm incomes would be defeated by free imports or any reasonable tariff rate.   Therefore, import quotas and other restrictions are frequently used to reduce the quantitative impact of foreign competition to a tolerable level.   Governments appear unwilling to tolerate an unrestricted policy of importation.

Import quotas impart a measure of certainty to economic affairs, especially in cases of inelastic and unknown foreign supply and they have the advantage over tariffs of administrative flexibility.   Nevertheless, problems do arise.   Fixing the quantity of imports is only the beginning.   How should the quota be divided among exporters?   Which importers should be permitted to use quotas and at which ports?   How should the quota be distributed over time?   How should the quota affect the income and balance-of-payments position of the country?   Is it likely to foster monopoly?   There is also the question of which units to allow—bushels or tons—and the problem of defining the products to be restricted.

In the previous chapter, it was shown that in the United States the utilization of import quotas has given more power to the administration, as opposed to the Congress, and in particular to administrators of trade policy.   It was also shown that import quotas subject trade to a larger number of arbitrary decisions on the part of government.   The difficulty of the search for an economic rationale was well illustrated in two reports, prepared by government agencies in the United States on the so-called "divided authority" problem in decision making in respect to import quotas.[5]

Of the principal reasons for the introduction and use of quotas—inelasticity in foreign supply, certainty and administrative flexibility—the most important is probably the certainty, or stability, which they bring about. This economic certainty has come at considerable cost to world trade, yet in many instances, countries regard import quotas as worth the cost.

Variable Import Levies

Variable import levies are classified as a form of import quota. The domestic price is usually fixed and is independent of the world price.[6] The levy is adjusted, in some cases, on a daily basis, to cover the margin between them. Theoretically, a variable import levy has the advantage that at least part of the margin of nominal protection accrues to the government as revenue.

The variable levy insulates the price structure of the countries which use it, so that while comparative advantage can to some extent influence the division of labor, it cannot operate to the advantage of producers outside. The variable import levy is commonly associated with the European Community although it originated in Sweden. It provides more protection than a tariff because it tends to reduce external competitors to the position of residual suppliers.

The variable levy is the perfect instrument for a policy of self-sufficiency; it acts like a quota inasmuch as it restricts imports and it denies foreign producers as a whole any increase in market share as a result of a reduction in world price. There are, however, important respects in which it differs from a quota.

First, quotas are often apportioned administratively among exporting countries, thus forestalling competition among them. The variable levy, on the other hand, is determined by the lowest import price, so that the lowest-cost exporters are still able to undersell the others and capture a greater market share. Second, whereas quotas fix the quantity of imports and cause domestic prices to absorb variations in domestic demand, variable levies shift any increase or decrease in demand to the quantity of imports in order to hold prices constant. Third, variable levies can be in principle, and have been, in practice, converted into import subsidies. In the mid-1970s world commodity prices rose higher than prices within the European Community for wheat, barley, corn, olive oil and sugar; importers therefore collected a negative levy from the Community budget. It is indeed this flexibility that chiefly distinguishes the variable levy from other methods of market insulation.

As originally conceived, the variable import levy used by the European Community was meant to foster greater efficiency in agricultural enterprises. In the long run, international comparative advantage would not have been distorted. To achieve this, however, prices should have been adjusted in the light of world supply conditions. In the case of the European Community, the result of the variable import levy together with the common agricultural policy has been the emergence of chronic surpluses of grains and dairy products which constitute a serious problem for the Community.

Health and Sanitary Regulations

The third category of nontariff measure that is of growing importance in agricultural trade is that of health and sanitary regulations. Article XX(b) of the GATT provides for the adoption and enforcement of phytosantiary and sanitary restrictions if they are "necessary to protect human, animal or plant life or health." All countries maintain health and sanitary regulations on imports as well as on domestic products. If the regulations apply with equal force to domestic as well as to imported products then those regulations cannot be considered as nontariff measures. On the other hand, if they are applied to imports in a discriminatory manner *vis-a-vis* domestic products, they would be trade-restrictive. Health and sanitary regulations (or phytosanitary and sanitary regulations) affect trade in livestock, meat, fruit and vegetables.

One of the problems with the enforcement of phytosanitary and sanitary regulations is the variation in standards among different countries. The CODEX Alimentarius Commission was set up jointly by the Food and Agriculture Organization and the World Health Organization in 1963 to develop internationally acceptable standards which would facilitate international trade while protecting the consumer at the same time. There are four main areas with which the CODEX Commission is concerned:

(a) the levels of pesticides, hormones and other residues in food;

(b) processing and packaging of food and food additives;

(c) labeling and nutritional information for the consumer;

(d) compliance with standards.

Not all countries have accepted the CODEX standards. In fact, one large area currently in dispute is concerned with the equivalence of different standards. Generally CODEX standards are based on microbiological and other product characteristics while in the United States food standards are based mainly on production and processing methods. This difference has led to a number of disputes between the United

States and the European Community. The United States is now press-
ing for recognition of the equivalence of their standards with those
based on microbiological product characteristics.

Most countries impose standards or specifications on domestic and
imported products in the interest of quality control and consumer protection. In the case of agricultural products, such regulations will be applied to fruit and vegetables. Insofar as such regulations are applied
without discrimination to domestic as well as to imported products then
they cannot be regarded as a restriction on trade.

A major shortcoming of Article XX(b), however, is that criteria
against which to measure the necessity of a restriction are not defined;
nor is a procedure delineated for settling disputes. This shortcoming in
this and other articles of the GATT led to the negotiation of the
Agreement on Technical Barriers to Trade (the Standards Code) during
the Tokyo Round of deliberations. The Standards Code was negotiated
by a limited number of nations, because in this way the difficulty of
arriving at consensus was minimized; the code is quite separate from
the articles of the GATT but countries, other than those among whom
the code was negotiated, may adopt its provisions.   The Standards
Code includes the obligation on central governments "whenever possible" to provide certification of standards by foreign testing agencies or
laboratories if the method of testing is approved. The implementation
of this obligation is evident in the recent agreement by Japan to accept
product certification by agriculture testing organizations in the United
States.[7]

Other Nontariff Measures

There are other nontariff measures which are used to restrict trade
in agricultural products. For example, imports of some products are
restricted during certain periods of the year, generally when domestic
production is being marketed. The restriction may be either a complete
prohibition or a limitation on quantity; it is usually applied to fresh
fruits and vegetables. A variation of this restriction is one where permission to import agricultural products of various kinds is dependent on
the sale of the domestic crop. In some countries, mixing regulations are
also used; this means that a certain prescribed proportion of domestic
output has to be used in such commodities as flour or tobacco products.

Measures to stimulate exports are often used in order to dispose of
excess agricultural supplies generated by high support prices. Both the
United States of America and the European Community have resorted
to export subsidies of this kind. Another form of subsidy to exports is
the requirement that agricultural exports should be shipped by national

flag carriers. Low interest loans are also often provided in industrialized countries.

The next section will examine the specific nontariff measures used in a number of countries.

Nontariff Measures in Current Use

At the end of the Kennedy Round of multilateral negotiations, a special committee was established charged with the task of finding mutually acceptable solutions to the problem of nontariff measures. A variety of reasons—economic, social, political, strategic, health and safety—were given by different countries to justify the imposition of regulatory measures. "Not unexpectedly, there was a tendency in many cases to explain that the trade effects of the measures were small. Also, not surprisingly, many were not convinced."[8]

From the detailed records compiled by the GATT Secretariat it has been possible to construct a list of the main nontariff restrictions by country and by type of product in 1969. The results are shown in Table 4.1. (The GATT Secretariat has made a number of attempts to update this information but it appears that there has been little change since 1969.)

TABLE 4.1   Summary of Agricultural Nontariff Barriers,
            Selected Countries, 1969

Country	Dairy Products	Grains	Livestock & Meat	Fruits & Vegetables	Vegetable Oils & Oilseeds	Tobacco	Cotton
Australia	HS	P	P	HS	HS	MR	SR
Austria	MIP SC QR L	MIP SC ST QR	MIP SC HS	QR SR Q	—	ST	—
Brazil	SC	ST BA QR	—	SC	SC	—	—

(continues)

TABLE 4.1 (*continued*)

Country	Dairy Products	Grains	Livestock & Meat	Fruits & Vegetables	Vegetable Oils & Oilseeds	Tobacco	Cotton
Canada	L	L ST	—	QR	—	—	—
Chile	ST SC	ST SC BA AD	—	—	AD	—	—
Denmark	L	MIP SC L	Q L HS	L Q SR	—	—	—
European Community	VL L	VL L	VL L QR	SR QR L MIP SC	VL L	QR ST	—
Finland	VL L	L ST	VL L	L ST	VL L	ST	—
Greece	—	L ST	—	L	—	ST L	SC
Ireland	QR	QR	QR	SR	—	QR	—
Israel	L ST	L ST	L	L	L	L	L
Jamaica	L	L	—	L	L	L	—
Japan	QR ST	ST	QR	QR	QR	ST	—

(*continues*)

48

TABLE 4.1 (*continued*)

Country	Dairy Products	Grains	Live-stock & Meat	Fruits & Vege-tables	Vegetable Oils & Oilseeds	Tobacco	Cotton
Korea	L	L	L	L HS	L	ST	—
New Zealand	—	ST HS	HS	BA ST	HS	MIP	—
Norway	QR	ST	QR	IC QR ST	ST	—	—
Peru	L SC	L SC	P L SC	P L SC	SC P L	SC P L	—
Portugal	QR	QR	QR	—	QE	ST	QR
South Africa	L QR	L QR ST	L	L ST	—	BA L	L QR
Spain	ST L	ST VL L	ST HS VL L	L	QR L ST	ST	ST
Sweden	MIP VL L	MIP VL	MIP VL HS L	MIP VL SR ST	ST VL	—	—
Switzerland	ST QR SC	MR SC ST	SC QR	SR QR	SC	—	—

(*continues*)

TABLE 4.1 (*continued*)

Country	Dairy Products	Grains	Live- stock & Meat	Fruits & Vege- tables	Vegetable Oils & Oilseeds	Tobacco	Cotton
United Kingdom	QR	MP SC	QR	QR	–	QR	–
United Kingdom	QR	QR$^{(a)}$	(b) HS	QR$^{(c)}$	–	–	QR

Source: Developed from information given in various circulars published by the Foreign Agricultural Service, USDA, and from unpublished data.

(a) On wheat and flour
(b) Automatic import controls imposed once a certain level of imports is reached
(c) On peanuts and sugar only

<u>Key</u>

AD Advance deposit required on imports
BA Bilateral agreement for the purchase or exchange of goods on special concession to facilitate imports
HS Health and sanitary regulations
L Licenses required before goods are imported
MIP Minimum import price restrictions
MR Mixing regulations—the finished product must contain a certain proportion of domestic output
P Imports prohibited
Q Quotas apply to imports
QR Quantitative restrictions on imports
SC Supplementary charges on imports—surcharges, special import taxes, etc.
SR Seasonal restrictions (sometimes referred to as import calendar). Restrictions on imports are imposed during certain periods when the domestic crop is being harvested and marketed.
ST State trading, that is, importation by a government or a government authorized agency
VL Variable levy

It will be seen from this table that quantitative restrictions and related licensing procedures are the nontariff measures most often identified. The type of quantitative restriction and variable levy used in different countries, however, varies considerably. In some cases levies are, in practice, varied infrequently and are thus similar in effect to a tariff, while others vary frequently and effectively equalize internal and import prices. The fact that two types of levy have the same *ad valorem* incidence, however, would not necessarily mean that they have the same effect. Each effort by the GATT Secretariat since 1964 to classify nontariff restriction has not changed the result noticeably.

The extent and general nature of nontariff measures applying to imports of agricultural and other primary commodities in the European Community, Sweden and Japan is shown in Table 4.2. This table, which has been adapted from material prepared by the UNCTAD indicates the general nature of the restraints imposed at national borders; it also shows the type of explicit government intervention which can occur in domestic markets.

TABLE 4.2   Nontariff Measures Applied to Imports of Agricultural Products, European Community, Japan and Sweden

Product	Type of Nontariff Measures Applied[a]		
	European Community	Japan	Sweden
Vegetable oils[b]	None	HS	V
Vegetables:			
Fresh or chilled vegetables	BQ,GQ,HS,OM,SR	HS	SR,V
Frozen vegetables	None	HS	None
Vegetables in brine	None	HS	None
Dried vegetables	BQ,Q,MIP	HS	None
Dried legumes	None	HS,DL,Q	V

TABLE 4.2 (*continued*)

Product	Type of Nontariff Measures Applied[a]		
	European Community	Japan	Sweden
Fruits:			
Tropical fruit	G,GQ,BQ,HS,OM	HS	None
Citrus fruit	HS,OM,GQ	HS,DL,Q	HS
Figs	HS	HS	None
Grapes	HS,SR,OM,DL	HS	HS,SR
Nuts	HS	HS,Q	None
Apples and pears	HS,OM	HS	L,HS,SR
Stoned fruit	HS,SR	HS	HS,R,SR
Berries	L,HS	HS	SR
Meat:			
Bovine animals	L,HS,OM,VL	HS,Q	L,VL,HS
Swine	VL	HS	L,VL,HS
Sheep	L,GQ,MP,OM	HS	L,VL,HS
Poultry	HS,OM,VL	HS	HS,VL
Fresh meat	DL,HS,OM,VL	HS,DL,Q	HS,VL,L
Poultry meat	MP,OM,VL	HS	HS,VL
Poultry liver	VL,HS	HS	HS,VL
Other meat	HS	HS	L,Q
Dairy Products:			
Fresh milk	OM,VL	DL,Q,HS	VL
Preserved milk	OM,VL	DL,Q,HS,ST	VL
Butter	OM,VL	Q,HS,ST	VL
Cheese	OM,HS,VL	DL,Q,HS, PLR,TQ	VL
Eggs	OM,VL	HS	VL,L
Cereals:			
Wheat	L,OM,VL	HS,Q,ST	VL,L
Rye	VL	HS	VL
Barley	L,VL	Q,HS,ST	VL
Oats	V	HS,Q	VL
Maize	L,MIP,OM,VL	HS,Q	VL
Rice	VL	Q,HS,ST	None

(*continues*)

TABLE 4.2 (*continued*)

Product	Type of Nontariff Measures Applied[a]		
	European Community	Japan	Sweden
Sugar and confection:			
Beet and sugar cane	VL	HS,MIP,TX	L,VL
Other sugars	VL	HS,TX,DL,Q	VL,L
Molasses	VL	HS,TX	VL,L
Sugar confection	BQ	HS,PLR,TX	VL

Source: Alexander J. Yeats. "Agricultural Protectionism: An Analysis of its International Economic Effects and Options for Institutional Reform," in *Trade and Development*, Winter 1981, pp. 12-13.

(a)   Restrictions applied in whole or in part to the group as classified by the Standard International Trade Classification.

(b)   Including soybean, groundnut, rape seed, cotton seed, coconut, palm and palm kernel oil.

Key:

BQ   Bilateral quota
DL   Discriminatory licensing scheme
GQ   Global quotas
HS   Health and sanitary regulations
L   Licensing of importers
MIP   Minimum import price restrictions
OM   Other price distorting restrictions
PLR   Special labeling requirements
Q   Quotas (method unspecified)
R   Unspecified restrictions
SR   Seasonal restrictions
ST   State trading
TQ   Tariff quota
TX   Special internal tax
VL   Variable levy

Incidence of Nontariff Measures

A noticeable aspect of the current discussion on nontariff measures in agricultural trade is the lack of hard evidence about the manner in which the regulations are applied, the degree to which trade is impeded by their application, and the magnitude of the protection afforded by the restrictions to particular commodities or commodity groups. The purpose of the GATT classification of nontariff measures and the compilation of the inventory was to assist in the exploration of ways in which distortions to trade arising from agricultural measures could be mitigated. So far there has been little quantification of the effect on trade in agricultural products of nontariff measures because of the difficulties of measurement.

The frequency with which nontariff measures are used varies widely among developed countries, reflecting in part their competitive position in world trade. It also partly reflects the degree to which the various countries rely on regulatory mechanisms in their trading. This is illustrated in Table 4.3 where import restrictions of all types are considered together. From this table it appears that France uses import restrictions to regulate its international trade to a far greater extent than either Canada or Australia. Of course, summaries and frequency charts are not measures of effective protection, although they are fairly reliable indicators of the potential harassment and restriction which could be caused by nontariff measures.

TABLE 4.3 Frequency of Import Restrictions on Selected Products[a,b]
Percent

Countries	Primary Commodities	Semi-processed Agricultural Products	Processed Agricultural Products	Semi-manufactures	Manufactures	Number of Products Affected
France	6.7	9.0	12.5	5.3	16.7	38
Federal Republic of Germany	3.9	6.9	10.8	6.8	—	31
Ireland	6.7	6.9	10.0	3.0	—	34

(*continues*)

TABLE 4.3 (*continued*)

Countries	Primary Commodities	Semi-processed Agricultural Products	Processed Agricultural Products	Semi-manufactures	Manufactures	Number of Products Affected
United Kingdom	5.6	6.9	11.7	2.3	—	31
Denmark	5.6	6.9	10.6	1.5	—	29
Italy	5.6	7.6	7.5	1.5	—	31
Benelux	5.0	6.9	7.5	2.3	—	30
Austria	10.0	9.0	7.5	—	—	23
Norway	8.9	6.3	2.5	—	—	21
Switzerland	7.8	4.9	5.8	—	—	16
Japan	3.3	4.2	5.0	—	—	18
United States of America	1.7	1.4	2.5	2.3	—	10
Finland	—	1.4	5.0	—	—	6
Canada	1.1	2.1	—	3.0	—	9
Sweden	—	4.9	4.2	—	—	5
Australia	1.1	—	0.8	—	—	3
Simple Average	4.6	5.3	6.5	1.8	1.0	

Source: The Processing Before Export of Primary Commodities: Areas for Further International Cooperation, Table 10V, TD/229/SOPP2 (Geneva: UNCTAD Secretariat, 1979).

(*continues*)

TABLE 4.3 (*continued*)

(a) Frequency is defined as the number of restrictions in a group as a percentage of the total number possible in a group. The denominator (the number of processing product group) × (12 types of restrictions).

(b) The table is made up of 49 products or product groups of 15 primary commodities, 23 processed and semi-processed agricultural products and semi-manufactures; and 11 manufactured products.

Measuring the frequency of restrictions by product provides another interesting perspective. Research conducted in the early 1970s indicates that of all processed agricultural products, cereal products are most subject to import restrictions in developed countries. At that time, the restrictions on cereal, meat and fruit products together accounted for two-thirds of the total.[9]

Measurement of Agricultural Protection

The need to measure the extent of agricultural protection is now recognized. International organizations have shown an interest in methods of measurement and in particular, the Food and Agriculture Organization (FAO) has attempted to develop a methodology for estimating the domestic and trade effects of government intervention in agricultural markets.[10] Studies of this kind should be the beginning of more serious attempts to measure the effects of farm programs and intervention in international agricultural markets.

Attempts have also been made to devise a theoretical model that could determine the effects of a nontariff barrier as a first approximation.[11] Two noteworthy models developed, respectively, by Larry Wipf and Alexander Yeats, use the effective protection (or effective tariff rate) concept developed in the analysis of the effect of tariffs and quotas. Unfortunately, this concept is inadequate for nontariff barriers that cannot be measured directly, such as health standards. Both authors use the effective rate of protection model in order to determine the effective tariff rate on various agricultural products in the United States and other industrialized countries. The difficulty of estimating the nominal rate of tariff equivalent for certain nontariff barriers is acknowledged, however, and it is evident that certain special assumptions were made in order to derive some of the estimates.

Three partial-equilibrium trade models explicitly illustrate the intuitive impact of enforcing a nontariff barrier, but the dearth of appropriate data to determine with reasonable accuracy the estimators of the demand and supply schedules is sadly evident.[12] Although they might accurately simulate the impact of nontariff barriers, the models are extremely difficult to test and to apply as a policymaking tool. The major problems are, first, to determine what a nontariff measure is and second, to assign a value to it. So far, no satisfactory model has been formulated.

Export Subsidies

In addition to restrictions on imports, countries use export subsidies and controls in an attempt to deal with problems of agricultural surpluses. This section will illustrate the type of trade-distorting measures used in the United States and the European Community but the measures are also used by other countries and trading blocks.

In the early 1950s, after a variety of control techniques had failed to stem farm production, the United States resorted to massive assistance to exports in the form of grants, soft loans to recipient countries and export subsidies. In some cases, the subsidy constituted a large percentage of the world price of the commodities exported.

Since 1972, a variety of measures designed to assist in the export of surplus production have been in operation. The measures include long-term sales on favorable terms and grant-type arrangements. While concessional sales accounted for only 15 percent ($1 billion) of the agricultural exports from the United States in the early 1970s, they were of great importance for certain commodities. In 1971, more than one-third of the wheat, a quarter of the cotton, half of the rice, more than 40 percent of the soybean oil and nearly 90 percent of the non-fat dried milk were exported under this plan. By 1973, the Commodity Credit Corporation held virtually no stocks, but nevertheless the legislation was extended for a further four years.

Commercial and quasi-commercial export schemes involve less subsidy than do concessional sales. In one of these export schemes, the Commodity Credit Corporation has extended short-term credit for a period from six months to three years to foreign importers of United States agricultural produce at interest rates low enough to be attractive to most countries which qualify for such credit.

Another type of commercial aid has been the cash subsidy made to exporters to enable them to meet day-to-day foreign competition in world markets. These payments have been made on wheat, flour, rice and tobacco. The commodities which the Commodity Credit

Corporation has acquired through price support arrangement or direct purchase have been made available to private commercial firms which then make arrangements to export the commodities. For example, the various grain deals between the United States and the Soviet Union have involved substantial government-owned stocks. Cash subsidies, favorable interest rates and subsidies to merchant shipping are also often part of such deals.

Structural surpluses are a more recent phenomenon in Europe. Wine lakes, butter and wheat mountains are evidence of a protected agricultural sector. As a complement to the variable import levy, the European Community pays export subsidies, "on restitutions," to facilitate the disposal of surplus agricultural products on world markets. Generally, the subsidy makes up the differences between prices prevailing in the European Community and world prices. Products are accumulated through support buying at intervention prices and exports are subsidized from the European Agricultural Guidance and Guarantee Fund within the Community budget; most of the funds for this body come from the import duties and value added taxes contributed by member countries.

Export Subsidies and the GATT

The status of export subsidies under the GATT is controversial. In all cases except for primary products their use is not permitted. Early in the history of the GATT the contracting parties had the opportunity to extend the prohibition to farm products. The United States was reluctant to do this, however, because the disposal of surpluses on the world market was a necessary outlet for the intervention purchases of the Commodity Credit Corporation. The special provisions for agriculture are stipulated in Article 16 of the GATT and also in the Subsidies Code adopted at the Tokyo Round of multilateral trade deliberations. These special provisions accept that export subsidies which harm the interests of contracting parties should be avoided always, but they are illegal only when they result in a "more than equitable market share" for the country that employs them. Parties whose exports are displaced are entitled to an investigation by a GATT panel and, following its judgment, may seek redress from the Council of the GATT.

The procedures in the GATT have rarely given sastisfaction. The law leaves too many issues open to interpretation. For example, it is not clear which products are classified as primary products or what constitutes a more than equitable market share. It is also difficult to separate the effects of subsidies from other influences on market shares and to know how much injury a complainant has to demonstrate in order to be entitled to redress.[13]

Moreover, it can be almost impossible for a complainant to secure a prompt favorable judgment and if the issue is referred to the Council of the GATT, one party has the power effectively to block any further proceedings. In such situations, injured parties find their own means of redress, but these often have the reverse effect. Retaliatory subsidies may force down the world prices and raise the cost of intervention to support domestic farm prices.

Multilateral trade negotiations will not make any progress towards resolving this issue while the deeper problem of which it is the symptom remains a domestic matter. As long as agricultural policies in developed countries continue to encourage over-production, there will be no end to the use of export subsidies in order to dispose of surplus production.

Export Controls

Another aspect of the same problem is the use of export controls. In the mid-1970s, a number of countries which exported agricultural products used export controls of various kinds to prevent world prices (which were increasing rapidly) from pushing up domestic consumer prices of grain and other food products. Such devices are not a novelty in world trade. Yet the cancellation by the United States of contracts for the export of soybeans in 1973, followed by the delay and partial cancellation of a grain sale to the Soviet Union in October 1974, received great publicity because they were anomalies; in the absence of routine controls and no constitutional authority to tax exports, the American government broke export contracts as an extraordinary resort. Export controls are an integral part of trade policy in some countries. Australia and Canada through their wheat and grain boards and the European Community through export taxes, licensing and embargoes control the export of a number of commodities. Other countries, too numerous to list, use a variety of devices to limit exports when necessary.

At the opposite extreme, export controls may be introduced where a country or countries intend to limit exports because of depressed world markets in the hope of increasing gross receipts for production and/or stocks. Tropical products such as coffee and cocoa have been subject to international commodity agreements with supply control being the fulcrum by which the economic objective is reached.[14]

There is a strong case against export controls in international trade. Just as import controls are questionable as an economic tool to provide help to a sick industry with weak international markets—the normal case of the past—export controls are a poor way to fight inflation. They present special dangers especially in their effect on the international value of the currency of the country imposing the controls. For example, if foreign traders cannot purchase grain from the United States, dollars become less valuable. In a world of floating exchange rates, that loss in value is soon reflected in what the dollar can buy and this will have an inflationary impact on the United States.

Another serious risk in restricting exports of vital commodities in world trade is that such restrictions inevitably distort patterns of world trade. Attempts to relieve domestic price pressures by restricting exports soon interferes with patterns in commodity trade, particularly if these limitations are imposed by a world power and major exporter like the United States.

The ultimate danger is that there could be a return to the conditions of the 1930s when international trade collapsed. The effect would be the same, whether trade is strangled by competitive import measures as it was in the 1930s, or by competitive export measures which was the threat during the mid-1970s. In the following chapter the extent to which some developed countries use nontariff measures to restrict trade in agricultural products will be examined in more detail.

Notes and Sources

1. *Nontariff Distortions of Trade* (New York: Committee for Economic Development, 1969), p. 12.

2. *Ibid.*

3. Among the FAS circulars in the Agricultural Trade Policy (ATP) series, see, for instance ATP-3-72 on Canada, ATP-8-72 on U.K., ATP-10-72 on the EC. These circulars were issued without specific titles; they were all published in 1972. They can be obtained from the U.S. Government Printing Office in Washington.

4. *Liberalization of Tariff and Nontariff Barriers.* Document TD/B/C.2/R.1 and Add. 1, (Geneva, UNCTAD Secretariat, 1969). Robert Hawkins and Ingo Walter, *The United States and International Markets: Commercial Policy Options in an Age of Controls* (Lexington, Massachusetts: Heath, 1972).

5. *Special Report on Examination of Divided Authority Over Agricultural Import Quotas Under Section 22 of the Agricultural Adjustment Act,* as amended, Comptroller-General of the United States (Washington: U.S. Government Printing Office: 1961); and *Study Group Review of the Special Report on Examination of Authority over*

Agricultural Agricultural Import Quotas Under Section 22 *of the Agricultural Adjustment Act,* as amended, USDA, October 1962, prepared in response to the Comptroller-General's report.

6. This assumption of independence becomes more difficult to justify in those instances, such as have developed from the common agricultural policy of the European Community, where supplies produced domestically have been a relatively large part of world supplies and/or a large part of domestic consumption. In this instance, intervention in domestic markets results in the necessity of dumping excess supplies on world markets.

7. I am indebted to Professor Maury E. Bredahl and Kenneth W. Forsythe of the University of Missouri-Columbia, for permission to draw on their article, "Harmonizing Phytosanitary and Sanitary Regulations," *World Economy,* Vol. 12, No. 2, June 1989, pp. 189-206.

8. *United States International Economic Policy in an Interdependent World,* Vol. 1, Section 5, Part 2, published for the Commission on International Trade and Investment Policy (Washington, D.C.: U.S. Government Printing Office, 1971), p. 621.

9. *Program for the Liberalization of Quantitative Restrictions and other Nontariff Barriers in Developed Countries on Products of Export Interest to Developing Countries,* TD/120/Supplement 1, (Geneva: UNCTAD Secretariat, 1972).

10. Timothy Josling and Jimmye Hillman, *Agricultural Protection and Stabilization Policies: A Framework of Measurement in the Context of Agricultural Adjustment,* C75/LIM/2. (Rome: Food and Agriculture Organization, 1975).

11. The following give more descriptive detail with regard to the theoretical models: Larry J. Wipf, "Tariffs, Nontariff, Distortions, and Effective Protection in U.S. Agriculture," *American Journal of Agricultural Economics, Vol.* 53, 1971, *pp.* 81-5; Alexander J. Yeats "Effective Protection for Processed Agricultural Commodities: A Comparison of Industrial Countries," *Journal of Economics and Business,* No. 29, Fall 1976, pp. 31-9; Hawkins and Walter, *op cit;* W.M. Corden, *Trade Policy and Economic Welfare* (London: Oxford University Press, 1974); and John Strack, *Measurement of Agricultural Protection* (London: Macmillan for the Trade Policy Research Centre, 1982). For an attempt to use the effective protection concept empirically, see Alexander Yeats, "Agricultural Protectionism: An Analysis of its Inter-national Economic Effects and Options for Institutional Reform," *Trade and Development: An UNCTAD Review,* (Geneva: Winter, 1981) pp. 1-29.

12. Robert L. Thompson, *A Survey of Recent U.S. Developments in International Agricultural Trade Models,* Bibliographies and Literature of Agriculture No. 21, USDA (Washington: Government Printing Office, 1981).

13. See Edmond McGovern, *International Trade Regulation* (Exeter: Globefield, 1986) pp. 328 and 450-1; and Beatrice Baudet, *The Agricultural Trade War* (Brussels: Bureau d'Informations Europeénnes, 1984), pp. 185-231.

14. The Cocoa Agreement which became effective on June 30, 1973 had little influence over the cocoa market. The purpose of the agreement was to stabilize cocoa bean prices within a given range by using export quotas in the six major producing countries. The quotas were reduced as prices fell. When prices approached the upper end of the range, sales were to be made from buffer stocks. Quotas were never implemented, however, as cocoa prices remained well above the price range and the buffer stock was not accumulated because supplies did not permit it.

5

Use of Nontariff Measures in Developed Countries

From the very large number of nontariff or technical measures available, some countries tend to rely heavily on one particular non-tariff measure—for example, the variable import levy in the European Community. No country, however, relies entirely on one protective instrument. Generally, a combination of measures are used to control trade in any given product. This chapter considers in more detail nontariff measures that are used by a number of developed countries.

United States of America

Conditions in agricultural production and trade in the United States have changed considerably in the past fifteen years, but the fundamental commitments to the farmer have not altered. As of 1990, all the basic farm and trade legislation and regulatory activities were still intact but a heavy reliance on exports made prices more volatile and unpredictable.

Table 5.1 presents the specific measures which have been used by the United States since 1973 to support prices of selected crop and livestock products. It illustrates the diversity of the protection available to legislators, administrators and regulators. Included in this table are domestic price supports and direct payments, production controls, export subsidies and tariff and nontariff import controls. Not included are the various indirect subsidies, such as conservation payments and subsidized credit. It should be noted that export subsidies, which played a significant role in American agricultural policy until 1972, were not important for the balance of that decade when prices of grain

63

and other agricultural products were high in world markets. Export subsidies were introduced again in 1984 when large grain surpluses had once again accumulated. (This trade-distorting practice is one of the principal measures addressed in the Uruguay Round of trade negotiations.)

Not all nontariff barriers in the United States relate to commodity price support programs. As has been shown in Chapters 3 and 4, there is a wide range of trade restrictions in the United States, as in most other countries, which have no direct ties to commodity price support programs—for example health and sanitary issues. In addition to the obvious animal disease-related restrictions such as *aftosa* or hoof-and-mouth disease there are various meat-inspection restrictions (these will be covered in more detail in Chapter 6). Imports of cured meat from the European Community have also long been subject to restrictive controls in the American market. For example, there has been a long-standing import prohibition on the import of Parma ham into the United States although negotiations in 1989 may have made importation of this product easier.

Similarly imports of horticultural and agricultural plants into the United States are subject to quarantine regulations. The Department of Agriculture oversees the administration of these regulations the purpose of which is to protect agriculture and livestock producers against importation of diseases and pests that do not exist in the United States. Over the past few years, these regulations have been the subject of discussion and negotiation between various countries, including the Netherlands, Belgium and Denmark. During this period, however, the United States has repeatedly postponed modification of the regulations, allegedly because of inadequate manpower to carry out the necessary scientific examination. Meanwhile, the Department of Agriculture is being subjected to strong pressure from American growers and producers not to amend the regulations. Some of the quarantine regulations are so restrictive as to bar imports from certain countries.

Other products, for example potatoes grown in Europe, and a large variety of plants from the Netherlands, Belgium and Denmark which use growing media such as rockwool, have been the subject of negotiation in recent years. The regulations covering the import of such products remain, however, inappropriately restrictive. The principal objective of Table 5.1 is to illustrate the diversity of the instruments of protection available to legislators and to administrators.

As can be seen from the table, however, the nontariff measure that is most frequently in use is the quota arrangement. The authority to impose restrictions of this kind in the United States is given by Section 22 of the Agricultural Adjustment Act which was approved by Congress

TABLE 5.1 Summary of Agricultural Support Measures in the
United States, 1990

Product	Price Support	Other Support Payments	Production Controls	Export Subsidy	Import Controls or Tariff	Nontariff Measures
Wheat	x	x	x	x	x	—
Feed grains[a]	x	x	—[b]	—	x	—
Rye	x	x	—	—	x	—
Rice	x	x	x	—	x	—
Dairy products	x	—	—	—	x	Q
Soybeans	x	x	—	—	—	—
Cotton seed	x	—	x	—	x	—
Peanuts	x	—	x	—	x	Q,QR
Beef and veal	—	—	—	—	x	Q,VER
Pork	—	—	—	—	—	—
Poultry	—	—	—	—	x	—
Tobacco	x	—	x	—	x	—
Sugar	x	—	x	—	x	Q

Sources: Economic Research Service and Foreign Agricultural Service, USDA, Washington.
(a) Including corn, sorghum, barley and oats. b) Voluntary set asides for corn, sorghum and barley and voluntary acreage reduction for corn, sorghum, barley and oats.
Key: Q = quota restrictions; QR = quality regulations; VER = voluntary export restraint.

in 1935.   This is one of the most important pieces of legislation
governing agricultural trade in the United States and it will now be
considered in more detail.

Section 22 of the Agricultural Adjustment Act

Section 22 has been amended several times since it was first
enacted; it has also been supplemented by additional legislation
concerning international trade agreements.   Because of its fundamental
importance in American trade policy and because of its implications for
protection, a complete text of Section 22 is presented in Appendix C.
Section 22 authorizes the President of the United States to restrict the
importation of commodities by the imposition of fees or quotas if such
importation would render ineffective or materially interfere with the
policies of the Department of Agriculture in relation to agricultural
commodities.   Section 22 requires the International Trade Commission
(originally called the Tariff Commission), on the direction of the President, to conduct an immediate investigation, including a public hearing,
and to make a report and recommendation to him.   Interestingly, no
time limit is set by which any such report has to be submitted.

The scope and permissible action of the original legislation was expanded by the Trade Agreements Extension Act of 1951, under which
no trade agreement or other international agreement can be applied in a
manner inconsistent with requirements found in Section 22.   The Trade
Expansion Act of 1962 also makes that exception.   One clause of Title
II of the latter Act reads as follows: "Nothing contained in this Act
shall be construed to affect in any way the provisions of Section 22 of
the Agricultural Adjustment Act, or to apply to any import restriction
heretofore or hereafter imposed under this legislation."

Special emergency measures to deal with imported perishable goods
were, however, introduced in the Trade Agreements Act of 1951.   Later,
Section 104 of the Trade Agreements Extension Act of 1953 clarified the
emergency action which could be taken with regard to the perishable
products; this is now incorporated in Section 22.   It allows the President
to take immediate action without waiting for the recommendation of
the International Trade Commission whenever the Secretary of Agriculture reports the existence of a situation requiring emergency
treatment.   The President is empowered to take immediate action and
any measure which is introduced to deal with the emergency may continue until the President receives the findings of the International Trade
Commission together with their recommendation.   The President may
take unlimited time to act on the recommendation, So far, however,
the emergency clause has rarely been used.

Three specific prerogatives are given to the President under Section 22, namely: (a) he may impose fees not in excess of 5 percent *ad valorem*; (b) he may reduce the importation, or warehouse withdrawal, of a particular commodity to no less than 50 percent of that of a representative period; and (c) he may, in designating any article or articles, describe them by physical qualities, value, use, or any other attribute.

Action under provisions of Section 22 is initiated in the Department of Agriculture, the primary responsibility being assigned to the Administrator of the Foreign Agricultural Service. From the first investigation in 1951 until 1990 there have been 51 investigations made by the International Trade Commission and its predecessor, the Tariff Commission. A large number of these were devoted to cotton and wheat and the latest two were on ice cream (1989) and on cotton-comber waste (1989–90). Cotton and wheat have, however, dominated Section 22 investigations because price support and price control policies have intervened significantly in the production and marketing of these products. Many of the later reports were concerned with the dairy industry —particularly with cheese. The major action was taken on cheese imports in April 1973 and again in January 1974.

Since the enactment of Section 22 in 1935, import controls have been imposed on twelve groups of commodities: wheat and wheat flour; rye, rye flour and meal; barley; oats; cotton, along with certain wastes and cotton products; certain dairy products; shelled almonds; shelled filberts; peanuts and oil; tung nuts and oil; flaxseed and linseed oil; sugars and syrups. Controls remain in operation until modified or revoked by presidential order.[1] By 1990 only four groups remained, those applying to cotton, dairy products, peanuts, and sugars and syrups.

The Section 22 quotas are maintained to protect the support program operating for the four commodity groups. Under the 1985 Food Security Act these programs combined all the major policy instruments used for the benefit of American agriculture—guaranteed prices, deficiency payments, diversion payments, production quotas and acreage diversion. Non-recourse loans set a floor for cotton prices and producers received a deficiency payment in addition to the sale price, which raised their returns to as much as, but not more than, a target price. To be eligible for support farmers could not plant more than 75 percent of acreage in upland cotton. It is unlikely that these programs of support will be substantially modified by the 1990 farm legislation.

The Commodity Credit Corporation supports the price of milk manufacturing through purchases of other milk products from processing plants. Under the original 1949 legislation, producer milk prices had to range from 75 to 90 percent of parity, a notion of "just price" which has a long history in the United States. This was accomplished by holding dairy imports to no more than one percent before sudden inflationary pressure made emergency quota increases necessary in the mid-1970s. The mounting surpluses of the late 1970s called for a freeze on levels of support and the prevailing prices in the United States dipped below 70 percent of parity. The government offered dairy farmers a number of incentives to reduce production, such as a herd buyout program; milk products were disposed of through social welfare and foreign assistance programs. Since the dairy glut was worldwide, tougher import restrictions were found to be necessary. The Trade Agreements Act of 1979 limited the quota on cheese imports for any succeeding year to 111,000 tons. By 1990 some adjustments had been made in quota applications and the quota total on cheeses from all countries were slightly higher at a little more than 113,000 tons. The world outlook for dairy products at the beginning of 1990 was considerably brighter and it is possible that there could be a further increase in the cheese quota under Section 22.

As for peanuts, non-recourse loans again set the floor for domestic market and the loan rate must rise in proportion to the index of prices farmers pay for production requisites. Support above the market price is restricted to production within stipulated quotas. Import quotas were stable over the decade 1980–90 at a negligible 1–2 million pounds (1.71 million pounds as of January 1, 1990). Sugar is the one commodity of the four of which the United States does not have a structural surplus. Although the non-recourse loan has tended to support prices at about three-to-four times the world prices, imports constitute some 30–40 percent of domestic supply. The motive for evading the quotas has evidently been strong and since 1983 the President has imposed emergency Section 22 quotas upon certain dry blends of sucrose and other sweeteners that were so constituted as to escape existing import restrictions.

Section 22 and the GATT

Although Section 22 was enacted in 1935, the provisions were not implemented until 1951 because of the prolonged disruption of world trade caused by World War II and its after effects. It was in that year that the United States Congress first imposed quotas on imports that threatened to interfere with domestic farm-support policies, notwithstanding international obligations to the contrary. More moderate and liberal import quotas were introduced in 1953, but at the insistence of the United States, exceptional treatment has been given in the GATT Article XI(2)(c) to quantitative import restrictions when they are combined with domestic supply-control measures. In 1955 the United States asked for and obtained a GATT waiver allowing it to use import quotas or fees as required under Section 22 to prevent material interference with agricultural policies in the United States. Other signatories to the GATT retained their right to compensation for any impairment of the concessions they had obtained from the United States.

The American position on agricultural trade liberalization has been compromised by the waiver and related trade restrictions. During the Kennedy Round negotiations, legislation that could be used to control imports of meat, sugar, processed dairy products and peanuts was being approved in Washington. Moreover, the substantial export-subsidy programs for agricultural products were maintained. Professor D. Gale Johnson afterwards wrote that "while farm groups and legislators from farm areas strongly supported reductions in barriers to trade in farm products, they had little appreciation of what changes had to be made in domestic policies to facilitate a liberal trade policy."[2]

Presidents of the United States and their Administrations have traditionally shown more appreciation and through Section 22 and related legislation the executive branch enjoys a commanding position in agricultural policymaking. The President, the Secretary of Agriculture and the International Trade Commission (ITC) are obliged to perform certain duties which invest them with wide-ranging powers. That these have been performed smoothly shows that the mechanics for hearing and acting upon a complaint are sound. It is moreover to the credit of the government of the United States that for the most part trade law is administered in the open air of public scrutiny and participation. Section 22 is no exception. Trade restrictions can be invoked only after full and complete hearings before the ITC at which evidence is welcomed from all interested parties, including foreign exporting interests.

Farmers in the United States have found Section 22 to be less restrictive in practice than it might have been. A particular example can be found in the cheddar cheese action of 1966. Secretary of Agriculture Orville Freeman infuriated the dairy industry by using the emergency provisions to increase the cheese quota. Furthermore, over the years there has been a substantial reduction in the number of products on which import restrictions, in the form of quotas, remain.

If Section 22 and related support programs that regulate trade have been administered with more procedural transparency and restraint than many instruments of protection, it is because protection is only one of the objectives for which this legislation was devised. The policy seeks to raise domestic prices above world prices for a large segment of the farm industry without seriously disrupting domestic resource utilization and historical patterns of international production and trade.

Needless to say, the internal contradictions of the policy have rendered this legislation unsatisfactory on one or all counts from the start. Domestic surpluses accumulate because price supports insulate the American market against foreign competition. And the practices involved are some of the same practices which the government of the United States has traditionally urged other governments to abandon in the interest of expanding international trade.

It will be shown in a later chapter that the United States as part of its negotiating position in the Uruguay Round of Multilateral trade negotiations, has expressed a willingness to negotiate the repeal of Section 22. If this piece of fundamental legislation were eliminated, and if no comparable substitute was put in place, the United States would, indeed, have taken a major step in removing the principal bulwark of its agricultural protection.

United Kingdom and the Commonwealth

As was seen in Chapter 2 the distinguishing feature of the United Kingdom's agricultural policy, from World War II until it entered the European Community, was the deficiency payments system. Under this system agricultural prices are determined by the free operation of the market but a minimum price is guaranteed to producers through direct payments for a specified quantity of their output. This is a convenient method of reconciling support for farmers with free entry for food imports. But it would be a mistake to label the system as non-protective. Dr. Timothy Josling has shown that the terms-of-trade effect of producer subsidies of this type, given the low price elasticities which exist for foodstuffs in general, restricts the volume of goods almost as effectively as an import quota or tariff.[3]

The United Kingdom became a member of the European Community on January 1, 1973 and on February 1 of that year the common agricultural policy was applied although a five-year transition period was allowed. One of the far-reaching implications of this move was a change in the countries which then had access to the UK market which traditionally had been a major outlet for a number of temperate-zone agricultural products.[4]

From 1973 to 1978, the United Kingdom was in a period of transition to the Community's system of variable import levies and other policy measures. During the transition period, the principal nontariff measures of the British market consisted of a minimum import price system, quota restrictions and prohibitions together with certain health, sanitary and quality controls. Table 5.2 presents a selected list of the nontariff restrictions which were used by the United Kingdom up to

TABLE 5.2 Selected United Kingdom Nontariff Measures, 1973

Commodity Description	Type of Restriction	Remarks
Fat cattle and fresh, chilled or frozen beef and veal, including offals	MIP, VL, HS	Foot-and-mouth and other disease controls. Offal not subject to MIP or VL.
Poultry meat and offals	MIP, VL, QR, HS	Bilateral quotas from Poland.[a] Newcastle and other disease controls.
Whole hams other than in air-tight containers	QR	Prohibited except from sterling area.
Fresh milk and cream	QR	Effective prohibition, except cream from Ireland.
Milk and cream preserved, concentrated, or sweetened	MIP, VL, QR	Licensing controls, except from sterling area (not restrictive). Bilateral quota for preserved cream from Poland.
Butter	QR	Bilateral quotas from traditional supplying countries.
Cheese	QR	Voluntary restraint agreements. Bilateral quotas from Poland.
Eggs	MIP	—
Bananas	QR	Global quotas from dollar area sources.
Fresh grapefruit	QR	Global quota from dollar area sources; bilateral quota from Cuba.

(continues)

72

TABLE 5.2 (*continued*)

Commodity Description	Type of Restriction	Remarks
Fresh apples and pears	QR	Global quota from all sources except sterling area. Apple quota divided into two periods with similar portion allocated to July/December period which is the main season from the United States.
Wheat and meslin	MIP, VL	The general variable levy/minimum import price system is the subject of agreement between the United Kingdom and the five major suppliers of grains to the United States. The agreement preserves the United States duty binding in the British market under GATT.
Clover and grass seeds	QR	Bilateral quotas from Poland. Quality controls.
Hop cones and lupulin	QR	Licensing controls, except from the Commonwealth.
Sausages	QR	Restricted from Poland.
Other prepared or preserved pork or poultry products	QR	Bilateral quota from Poland.
Beet sugar and cane sugar	QR	Quotas reserved for Commonwealth Sugar Agreement countries.

(*continues*)

TABLE 5.2 (*continued*)

Commodity Description	Type of Restriction	Remarks
Fruit jams and jellies	QR	Global quota from dollar area sources.
Grapefruit and orange juice, except frozen concentrates; apples	QR	Global quota from dollar sources.
Black currants and pear juice and syrup	QR	Bilateral quota from Poland.
Animal feedstuffs containing more than 80 percent by weight of milk solids	QR	Licensing controls, except from sterling.

Source: Compiled from various circulars and unpublished data of the Foreign Agricultural Service, USDA, 1972.

(a) Although this table includes quota restrictions maintained against Poland, which is a member of the GATT, it does not include those maintained against other East European countries which are not parties to the GATT.

Key:
HS   Health and sanitary regulations
MIP   Minimum import price restrictions
VL   Variable levy
QR   Quantitative restrictions on import

1973. Even though the United Kingdom is now a member of the European Community, many of these restrictions are still operative.

The table shows that minimum import prices existed for fat cattle, beef and veal, milk powders, preserved or condensed milk and cream (including animal feed containing milk solids), poultry meat, eggs, wheat, barley, oats, corn, grain, sorghum, cereal flour and meals and bran. A target price was established for each of these products, then a minimum price set a level high enough to prevent imports from entering the market at substantially less than the target price. Variable levies were used in most cases to ensure that the import price did not fall below the established minimum. This system of support was a move away from direct deficiency payments and made possible a smoother adaptation to the common agricultural policy.

For wheat, butter, cheese and bacon the United Kingdom steadily moved towards market-sharing and quantitative regulation even before entering the Community. To maintain a reasonable price in the case of bacon imports, the United Kingdom allocated a minimum quantity plus a reserve amount among her suppliers, including domestic producers. For butter, import quotas were allotted to overseas suppliers and quotas on butter and cheese safeguarded the pricing and marketing system for fluid milk. Until mid-1975, quota arrangements through the Commonwealth Sugar Agreement protected commonwealth suppliers by fixing the quantities of sugar to be imported with prices renegotiated every three years. This Agreement has now terminated and sugar imports now come under the regulations of the common agricultural policy.

The health, sanitary and quality control regulations in the United Kingdom are not entirely typical of all countries. Disease control measures are used to protect the domestic livestock industry, such as the ban on imports of pork and pork products from countries in which hog cholera exists. Fat cattle, beef and veal imports are subject to rigorous licensing requirements in order to control foot-and-mouth disease. In addition, imports of red meat are subject to a regulation prohibiting additives which maintain color. Strict quarantine, inspection and licensing conditions are maintained on shipments of live horses.

There are stringent licensing requirements for imports of dead poultry and offal when packaged separately and licenses are issued only for poultry which can be shown to be free of Newcastle disease. Other import requirements apply to dressed poultry and offal. A further quality control requires the licensing of clover and grass seeds so that imports meet further minimum standards and are appropriate for British soil and climatic conditions. The United States, in particular, regards import constraints by the United Kingdom on individual commodities as a limitation on the market for its agricultural products.[5]

Australia, Canada, New Zealand and other Commonwealth countries faced certain drastic trade adjustments with the entry of the UK into the European Community. New Zealand, for example, had to look to other markets in the Pacific and Middle East as it was gradually eased away from the captive market it had enjoyed in the United Kingdom. Other countries which had enjoyed a similar economic association through preferential agreements were also affected. Trade relationships between the United Kingdom and the United States which have existed for over a century underwent further changes when the full impact of the common agricultural policy was felt. For some countries in the British Commonwealth these arrangements have had an importance comparable to the abolition of the Corn Laws in the nineteenth century.

British policy dominated the agricultural trade of the Commonwealth countries until the late 1960s. During that time a number of them, particularly Australia, Canada and New Zealand developed measures to protect themselves against world price fluctuations and also to counter protective measures taken by other countries. The agricultural marketing boards of Canada, Australia, New Zealand and South Africa provide the best examples of government intervention in agriculture and cooperation with producers to raise incomes in the rural sector— usually through assistance. The main aim of these boards is to control the export of particular products. The six most important products with respect to export control are deciduous and citrus fruits in the Republic of South Africa; dairy products, apples and pears in New Zealand; and dairy products and wheat in Australia. All of these boards have export monopoly powers except the Australian Wheat Board, which controls export licensing. The "balanced development" policies of Australia entailed extensive support measures for farmers including subsidies, quotas, production controls, stringent health and sanitary regulations and various stabilization schemes.

In Australia, the importation of a wide range of products have been prohibited or severely curtailed because of strict plant or animal health regulations. Livestock imports are limited to breeding animals and very little meat is imported. Quantitative regulations mean that many horticultural products cannot be brought into the country. For example, apples and pears are prohibited because of the possibility of introducing the fireblight virus into Australia; and the import of avocado pears is prohibited unless they are certified as being free of sun blotch and blackstreak.

In May 1988, however, the Australian government announced a comprehensive liberalization program for agricultural products which should make their producers more efficient and competitive. The program includes reductions in agricultural tariffs, subsidies, government underwriting assistance and border protection in the general sectors, all to be phased in over a four-year period.

Canada, whose principal export is wheat, established a wheat board in 1935. As a government trading monopoly, it has dominated that country's international trade policy in wheat. Other products, in particular dairy products and grains, are subject to import permits. Approximately half of Canada's agricultural imports are duty-free, while tariffs on the remainder averaged about eight percent in the 1970s. On occasion, export subsidies have been paid on dairy products.

After much discussion, the United States and Canada signed, on January 2, 1985, a comprehensive free trade agreement aimed at the ultimate elimination of most trade barriers between the two countries. This agreement, which came into force on January 1, 1989, is of great importance because of the large volume of agricultural trade between the two countries: approximately US $4 billion in 1987 (about US $2 billion each way).

European Community

As Western Europe recovered from World War II, there were substantial increases in agricultural production but, although some progress had been made in removing quantitative restrictions on trade in agricultural products among Community countries, a large number of products remained subject to state intervention and control.

The first steps towards a common agricultural policy for the European Community were taken at a conference at Stresa in 1958. This led to a set of Community regulations and by the late 1960s a large percentage of agricultural production in the six countries came under the common agricultural policy.

Successive expansions of the Community from six members to nine, then to ten and, in 1986 to twelve countries make questions about protection more and more important. Table 5.3 illustrates the extent to which agriculture is protected under this policy, showing the protective measures used in 1985. The wide range of nontariff measures make markets for agricultural commodities in the Community some of the most highly protected in the world.[6]

For a number of years after 1972, the level of protection for agriculture in the Community diminished and the question arose whether this was a permanent trend or merely a reflection of conditions in the world economy. Doubts were soon resolved, however, for the late 1970s and early 1980s saw increases in the level of nontariff restrictions within the Community. The enlargement of membership in 1979 and again in 1986 is likely to enhance protectionist tendencies. Even though the common agricultural policy has been under considerable financial strain and criticism in the last few years, it has not undergone any fundamental change.

A so-called "reference" price, based on market prices in the Community is used to calculate compensatory taxes for various products. Imports found to be selling at less than the reference price may be subject to an offsetting tax. This tax may vary seasonally, as in the

TABLE 5.3 Summary of Agricultural Support Measures in the European Community, 1988

Product	Target Price	Inter-vention	Produc-tion Control	Export Subsidy	Import Levy	Non-tariff Mea-sures
Wheat	x	x	x	x	x	L
Feed grains[a]	x	x	—	x^b	x	L
Rice	x	x	—	x	x	L
Dairy products	x	x	x	x	x	L
Soybeans	x	x	—	—	—	—
Sunflower seed	x	x	x	—	—	—
Rapeseed	x	x	x	x	—	—
Cottonseed	x	—	—	—	—	S
Beef	x	x	—	x	x	L
Pork	—	x	—	x	x	L
Poultry	x	x	—	x	x	—
Tobacco	x	x	—	x	x	L
White sugar	x	x	x	x	x	L

Source: *Trade Policies and Market Opportunities for U.S. Farm Exports*, 1988 *Annual Report*, Foreign Agricultural Service, USDA (Washington: Government Printing Office, 1989).

(a) Including corn, sorghums, barley, oats, rye.
(b) Export subsidy on barley and rye.
Key:
L   License required to import;
S   Subsidy given.

case of fresh fruits and vegetables. It applies to fresh apples, cherries, grapes, lemons, oranges, peaches, pears, plums, tomatoes, wine and other minor products. Regulations provide for the imposition of certain types of compensatory taxes on imported oilseed and oilseed products as well as vegetable oils, subject to conditions surrounding their export[7] and with reference to their substitution for domestic products in the Community markets.

As in most countries, import licenses are used widely as a regulatory instrument in the Community's import control system. Licenses are required for all imports of grain, rice, olive oil, sugar, dairy products, frozen beef and veal and wine. Licenses are also required for imports of processed fruits and vegetables containing added sugar when the importer requests that the levy be fixed in advance.

In addition to the instruments which affect imports directly, there are a number of subsidies which make domestic products more competitive. There are also many administrative rules and regulations which, in practice, tend to restrict agricultural imports, many relating to quality standards.

The members of the Community apply their own health and sanitary regulations on most agricultural products (but see Chapter 6 with reference to red meat). This prerogative has resulted in numerous discrepancies in the standard for high value products which inhibit trade flows not only into the Community from outside suppliers but also among member countries; such discrepancies offer the cover of legitimacy to arbitrary and discriminatory restrictive practices. In Italy there is a prohibition on health and sanitary grounds on imported fruit, with few exceptions. Germany's standards on food and other agricultural imports are more stringent than are those of most other countries in the Community. Its packaging regulations limit fruit and vegetable juices to only five container sizes and peculiar labeling requirements are also imposed. In France, where administrative controls are notorious for their stringency, imported honey must be accompanied by documents certifying that it was produced in an area free of nocema, a disease that already exists in France.

No significant liberalization of nontariff measures has been made by the European Community as a whole in recent years. Nor, with the exception of the Netherlands, have individual member countries relaxed their own nontariff restrictions. The Community has reduced tariffs on certain commodities from the countries with which it has negotiated preferential trade agreements, but few reductions have been made to non-preferred suppliers. Coupled with sharp increases in the level of price support on many products, especially grains and rice, nontariff trade measures in the European Community constitute formidable obstacles to a more liberal trade policy in agricultural products.

At the beginning of the 1990s, some agricultural trade between the United States and the European Community is at risk because of disputes over health and sanitary issues in the meat trade. The first concerns the prohibition by the European Community, scheduled to take effect from January 1, 1989, of the importation of meat from any country which permits the use of hormones in livestock production for anything other than therapeutic purposes. This is likely to disrupt American exports of meat and edible offals to the Community. The second issue is the adoption of a directive by the member states in the Community which establishes minimum health standards for meat packing establishments that provide meat for sale in the European Community. These standards, which took effect on April 1, 1988, prohibit certain practices which are common in the United States. Because relatively few American plants have completed the expensive modifications necessary in order for them to be certified to export to the Community, the directive limits the access of products to that market. Producers whose plants have been certified, however, have maintained trade at pre-April 1 levels. The United States Foreign Agricultural Service continues to monitor the situation.[8]

It has been suggested that measurement of the increases in protection brought about by the common agricultural policy should be calculated by converting total charges of the variable import levy at the Community's border into *ad valorem* tariff equivalents.[9] This calculation shows that, for live animals, protection rose from 14 percent in 1959 to 49 percent in 1968; for meat, from 19 percent to 52 percent; and, for dairy products, from 19 percent to 137 percent. Protection for butter rose from 30 percent to 35 percent during this period and protection for cereals rose five times from around 14 percent before the common agricultural policy was introduced to 72 percent in 1968. Table 5.4 represents another approach to measurement of the impact of the variable levy. It traces the trend from 1960 to 1980 in the percentage difference between the Community and world prices for major commodities. If the figures are somewhat lower than expected, it is because the denominator of the support ratio is, in these calculations, the domestic rather than world price level. Commodity price inflation during the second decade (1970-80) covered by the table eroded some of the increase in the differential occurring in the first decade.

TABLE 5.4 Price Support Measures for the European Community, 1960-80 (Percent)

	1960	1965	1970	1975	1980
Wheat	27	29	35	3	15
Barley	24	16	40	7	19
Rice	28	26	28	8	30
Grains (including rye, oats, corn)	22	25	32	5	19
Beef	38	41	43	38	48
Pork	23	30	17	16	12
Chicken	26	25	6	−3	2
Eggs	20	18	13	7	6
Milk	23	30	46	37	35
Livestock products	27	32	34	28	29
Sugar beet	50	46	48	−8	29
European Community Total	24	28	32	21	26

Source: Yujiro Hayami and Masayoshi Honma, *The Agricultural Protection Level of Japan in an International Comparative Perspective*, Research Report No. 1 (Tokyo: Forum for Policy Innovation, 1983) p. 20.

Japan's Agricultural Policy

Japan became the world's largest importer of agricultural products by the beginning of the 1980s. She imported goods worth $15.8 billion in 1981. But for several years before, other nations had been observing Japan's trade policy with great interest. The choice of Tokyo as the site for the meeting of the council of GATT ministers in 1973 heralded Japan's growing importance in world trade and trade negotiations.

The history of Japan's agricultural protection has not been as well documented as has that of Western Europe and the United States. Japan was of relatively minor importance as an export market before World War II and its domestic farm and agricultural trade policies were of little concern to Western nations. In the post-war period of economic rehabilitation from 1946 to 1954 producers won ownership rights to their land, but the growth and productivity of the industry was encumbered by price controls, rent controls, and farmsize restrictions. Then rapid growth in the manufacturing and service sectors shifted the nation's comparative advantage relentlessly against agriculture and created a structural export surplus that threatened to upset the harmony of Japan's principal bilateral trading relations. For these reasons, during the two decades of the "economic miracle," the Japanese government sought to open the domestic market to foreign foods and raw materials.

In 1955 Japan became a contracting party to the GATT and since then tariff duties on many agricultural products have been reduced or eliminated.[10] With respect to Japan's trade with the United States, its largest supplier, tariffs remain on feedgrains and oilseeds as well as on cotton, hides and skins, inedible tallow, lemons, raisins, and almonds although these tariffs have been reduced. The process of step-by-step liberalization has included the relaxation of many nontariff restrictions as well. Many quotas have been removed or enlarged and in 1984, quota restrictions applied to 22 agricultural products (Table 5.5). Import quotas are usually combined with import licensing, however, as well as other administrative controls such as prior deposits, credit restrictions and other administrative or supplementary charges.

Nevertheless, the revolution in the diet of the Japanese people has been accommodated through these measures and it has been estimated that agricultural imports increased at an annual average rate of 7.9 percent during the period 1960-84. Self-sufficiency ratios have fallen for all grains and feedstuffs except rice as well as for many other commodities. By the 1980s, Japanese food self-sufficiency in terms of original calories was less than 50 percent, compared with nearly 80 percent in 1960.[11] The instability of world agricultural production and

82

TABLE 5.5   Agricultural Items under Residual Import Restrictions,
            Japan, 1984

Dairy products	Milk, fresh cream Evaporated milk, etc. Processed cheese, etc.
Meat and processed meat	Beef Beef products, pork products
Processed rice and wheat	Rice flour, wheat flour Rice meal, wheat meal, crushed feed Barley
Fruits, vegetables and processed products	Oranges (fresh) Oranges (stored) Fruit juice, tomato juice Fruit puree, fruit paste Processed pineapples Tomato ketchup, tomato sauce
Starch and sugar	Glucose, lactose Starch
Regional farm products and seaweeds	Beans Peanuts (excluding those for use in oil) *Konnyaku imo*, edible seaweeds
Marine products	Herring, cod (perishable or frozen) Herring, cod (salted or dried) Scallops, scallop eyes, cuttlefish
Others	Processed foods

Source: Industrial Review of Japan, 1984 (Tokyo: Nihon Keizai
Shimbun, 1984).

trade during the 1970s, however, revealed to the government and to the public the danger of this trend. To enhance the nation's food security the government revised its long term strategy for domestic farming and for trade liberalization. The comprehensive statement of policy for the 1980s, the *Basic Direction of Agricultural Policy*, reveals a new ambivalence about accommodating international comparative advantage. Reluctantly the government continues to make piecemeal concessions to freer trade, while at the same time it does not encourage—even if it does not actually resist—any reforms that would disturb the structure of the farm sector or sacrifice domestic market share.

Table 5.6 demonstrates that, in spite of a move towards trade liberalization, many nontariff trade restrictions remain as adjuncts to a highly interventionist farm price support program. Even though this table relates specifically to agricultural imports from the United States the restrictions are representative of Japan's trade policies toward all exporting countries. Many of these restrictions rely heavily on administrative interpretation—which can be affected by factors other than the letter of the law—so that an element of uncertainty is added to long-term prospects of foreign trade. These nontariff measures constitute a formidable wall of restrictions on international trade.

In recent years the agricultural protection in Japan has received a good deal of attention from countries which export agricultural products. Pressure from foreign commercial firms and governments has been reinforced by the many recent scholarly studies undertaken to reveal how and for what purposes this recondite system works.[12] In Japan, however, despite its costliness, agricultural protection has managed to command broad public support. Policymakers have succeeded in selling a sectoral policy to the public as a national food security policy that apparently serves the interests of everyone; non-farm interests in Japan have brought little pressure to bear upon the government to revise and reduce the protection of agriculture. Japanese food trade barriers have recently become quite controversial and important trade relations, notably with the United States, have been seriously upset. In spite of this, however, concessions to trade partners and reform of the protective structure are slow because, when it comes to agriculture, the international community has never been able to agree on a regime governing the practices of individual states.

Exports of agricultural products to Japan have been greatly impeded by the health and sanitary requirements of that country. Horticultural products bear the brunt of the regulations. For example, many of these deal with the codling moth and the importation of a substantial number of products including apples, pears, peaches and plums which are subject to this pest, is prohibited. Codling moth problems

TABLE 5.6 Japanese Trade Practices Impeding Agricultural Imports
from the United States, 1988

Commodity	Policy or Practice	Reason for Trade Measure
Feed grains	Restrictions on feed compounding outside bonded mills	Protection of feed industry
Corn	Tariff quota on corn for industrial use	Protection of potato starch industry
Rice	State trading	Protection of producers
Vegetable oils	Higher tariff imposed on unrefined oils with low acid content	Protection of refiners
In-shell peanuts	Plant quarantine restrictions	Protection of producers
Peanuts	Import quota; cartel marketing policies	Protection of producers
Beef cattle	Lack of quarantine facilities; tariff quota; high tariff on fat cattle	Protection of producers; budget constraints
Pork, fresh, chilled, frozen	Variable levy	Protection of domestic industry
Poultry, especially boneless & further processed	Tariff	Protection of domestic industry
Eggs	Tariff	Protection of producers
Canned fruit	Ban on use of sodium benzoate as preservative	Health protection

Source: Trade Policies and Market Opportunities for U.S. Farm Exports, 1988 Annual Report, Foreign Agricultural Service, USDA (Washington: Government Printing Office, 1989).

relating to walnuts, cherries, and nectarines were resolved when, after long and costly studies, the fruit and vegetable producers in the United States demonstrated that fumigation with methyl bromide is effective in killing the moth. Treatment with methyl bromide has been accepted by Japan for import of nectarines and cherries on a seasonal basis. Restrictions will be removed by 1991 for nectarines and by 1992 for cherries. It is hoped that restrictions on the importation of other fruit, such as apples, will follow shortly.

Another health and sanitary regulation in Japan concerns the importation of live cattle, particularly cattle imported for slaughter. For many years Japan has been unwilling to provide the amount of space required to implement the quarantine restrictions which are in force. Because of this, an artifical limit is imposed on the import of cattle into Japan. (This point is dealt with more fully in the following chapter.)

Also in the phytosanitary area of trade restrictions there is Japan's policy on food additives. The policy of the Ministry of Health and Welfare is highly restrictive. For example, many food additives which are commonly used in the United States, such as Red 40 food coloring, cannot be used in Japan, and foods containing such additives cannot be imported. In some cases, exporting companies have to be able to reformulate products for the Japanese market, but in many cases exports are impossible because such reformulation is not economically or technically feasible.

Japan's so-called "administrative guidance" system deserves particular comment, because in the panoply of protectionist instruments used in developed countries it is unusual. Government bureaucracy exerts its influence over business firms through administrative signals and suggestions. This is discussed more in the following chapter.

Notes and Sources

1. "Import Controls under Section 22 of the Agricultural Adjustment Act, as Amended," mimeograph, USDA, February 1981; "Report of the US Government to the Contracting Parties on Action under Section 22 of the Agricultural Adjustment Act." Trade Policy Staff Committee, USDA, November 29, 1983.

2. D. Gale Johnson, *World Agriculture in Disarray* (London: Fontana-Collins, 1973), p. 25.

3. T.E. Josling, *Agriculture and Britain's Trade Policy Dilemma*, Thames Essay No. 2 (London: Trade Policy Research Centre, 1970).

4. *History of Agricultural Price-Support and Adjustment Programs*, 1983-84: *Background for* 1985 *Legislation*, Economic Research Service, USDA, 1985. For more information on the general subject of the United Kingdom and the Community see Dennis K. Britton, "Problems of Adapting U.K. Institutions in the EEC," in John Ashton and John Rogers (eds) *Reforming the CAP* (London: Heath, for the Agricultural Adjustment Unit, University of Newcastle-upon-Tyne, 1973).

5. *Trade Policies and Market Opportunities for US Farm Exports*, 1988 *Annual Report*, Foreign Agricultural Service, USDA, 1989.

6. *Annual Review and Determination of Guarantees*, Cmnd. 3965 (London: HM Stationery Office, 1969), p. 5.

7. For example, imports of marine or vegetable oils may be subject to compensatory tax if the European Community finds evidence that such products receive a subsidy from the exporting country.

8. *Trade Policies and Market Opportunities for US Farm Exports*, op. cit., pp. 76-9.

9. See Harald Malmgren and D.L. Schlechty, "Rationalizing World Agricultural Trade," *Journal of World Trade Law*, No. 4, July-August, 1970, pp. 515-47.

10. For a summary of Japan's trade policy and protection in the period of transition before the Tokyo Round, see *International Agricultural Adjustment*: *A Case Study of Japan*, C73/LIM/3 (Rome: Food and Agriculture Organization, 1973), pp. 24-8.

11. *Japanese Agricultural Policies*: *Their Origins, Nature, and Effects on Production and Trade* (Canberra: Australian Bureau of Agricultural Economics, 1981), appendix.

12. See Kym Anderson and Yujiro Hayami, *The Political Economy of Agricultural Protection*: *East Asia in International Perspective*, (Sydney: Allen and Unwin, 1986); Jimmye S. Hillman and Robert A. Rothenberg, *Agricultural Trade and Protection in Japan*, Thames Essay No. 52 (Aldershot, England: Gower for the Trade Policy Research Centre, 1988), which contains an extensive bibliography on the subject of Japanese agricultural and trade policies; Fred Sanderson, *Agricultural Protectionism*: *Japan, United States and the European Community* (Washington: Brookings Institution, 1983).

6

Nontariff Barriers to Trade
in Meat and Livestock

In the earlier chapters, a brief outline was given of the nontariff measures that stand in the way of liberalization of trade in agricultural products. One of the most effective of these is the use of phytosanitary and sanitary regulations which, as pointed out in Chapter 4, are concerned with protecting the health of consumers and in creating an environment free of disease in both animals and plants. They are also, however, used as a method of trade protection as will be shown in what follows. This chapter deals in some detail with health and sanitary regulations in relation to trade in meat and livestock but other nontariff measures which impede or distort in these products are also considered and some examples of barriers to trade are given. Trade in meat and livestock has been selected because of the effectiveness of the health and sanitary measures that are used and also because of the diversity of other nontariff measures that are, or have been applied to this trade.

Trends in Production and Trade

World production and trade in all types of meat has increased markedly since World War II under the influence of income growth in both industrial and developing countries. In the period 1964–82 the volume of trade in red meat rose by 135 percent. The largest component of this increase has been trade in beef which doubled between the early 1960s and the late 1970s. From the mid-1970s, however, the consumption of beef in many OECD countries began to decline. This shift can be attributed not only to the rise in the price of beef relative to other meat but also to a change in tastes activated by an awareness of the risks that red meats (beef, pork and lamb) pose to health.

Beef producers in the major trading economies such as the European Community, the United States of America and Australia, were slow to adjust to the change in demand. In the European Community production had greatly expanded because of the price incentives of the common agricultural policy; by 1973 large surpluses had accumulated. Since then, supply has continued to outstrip demand and within the Community stockpiles have grown. To dispose of the surplus production, it became, in the 1980s, the world's second largest beef exporter.

The European Community is not alone in fostering domestic production of beef while at the same time maintaining prices and restricting imports. Similar measures have been taken in Japan and both the United States and Canada have adopted the same kind of tactics. Nontariff measures of all kinds have been brought into use but they have not resolved the structural problems of the livestock and meat sector.

An Overview of Trade Restraints

The restrictions used in trade in meat and livestock include tariffs as well as a variety of nontariff measures such as (i) specified dyes for marking carcasses (ii) the use of veterinary services and meat inspectors (iii) specified shipping documents and (iv) health and sanitary rules and regulations.

In Table 6.1 the types of nontariff measures that are used (in addition to any tariff) to restrict imports of red meat are listed while Table 6.2 shows the measures used to promote meat exports. The importance of each of the measures will vary from country to country. In Australia and New Zealand, for example, imports of beef are controlled through the direct application of health standard requirements while in Japan, in addition to rigid sanitary requirements, all beef imports and purchases of cattle have to be made through a quasi-governmental institution.

These requirements tend to restrict access to markets and to reduce the comparative advantage which a producer may have otherwise. Governments have numerous goals which they hope to achieve with their domestic agricultural policies; these goals include the achievement or preservation of self-sufficiency or the stabilization of prices and income in the agriculture industry.

TABLE 6.1 Nontariff Measures Applied to Meat Imports

Type of Measure	European Comm.	United States	Japan	Canada	New Zealand	Australia
Quotas	x	x	x	x	—	x
Bilateral quota	x	x	x	x	—	x
Tariff quotas	x	—	x	—	—	—
Voluntary export-restraints	x	x	—	—	—	—
Import licensing or permit	x	x	x	x	x	x
Variable levy	x	—	x	—	—	—
Countervailing duties	x	x	—	—	—	—
Minimum import price	x	—	x	—	—	—
Antidumping duties	—	x	—	—	—	—
Arbitrary	—	x	—	—	—	—
Cyclical control	—	x	—	—	—	—
Trigger quota	—	x	—	—	—	—
Release clause	—	x	—	—	—	—
Safeguard measure	x	x	x	—	—	—
State trading	—	—	x	x	—	—
Quasi-government institution	(a)	—	x	x	—	x
Special labeling requirements	x	x	x	x	x	x

(*continues*)

TABLE 6.1 (*continued*)

Type of Measure	European Comm.	United States	Japan	Canada	New Zealand	Australia
Advertising restrictions	(a)	—	—	—	—	—
Prohibition for health & sanitary reasons	x	x	x	x	x	x
Health & sanitary regulations	x	x	x	x	x	x
Tech. regulations	x	x	x	x	—	x
Packaging stds.	x	—	x	x	—	x
Grading differentiation	x	x	x	x	—	—
Meat cuts (differentiation)	x	x	x	x	—	—
Product differentiation by source	(a)	x	x	—	—	—
Influence of lobbies	x	x	x	x	x	x
Industry and labor	x	x	x	x	x	x
Consumers	weak	x	—	—	—	—

Source: Mark B. Lynham, *Nontariff Agricultural Trade Barriers: Livestock and Meat Legislative Regulation Devices as they Affect International Trade Between Industrially Developed Countries.* Unpublished Master's Thesis, Department of Agricultural Economics, University of Arizona, 1983.

Key —   indicates that no measures of this kind are applied or that insufficient information was available to determine whether such measures are used.

(a)   some countries in the Community, not all.

TABLE 6.2 Measures Used to Promote Meat Exports

Type of Measure	European Comm.	United States	Japan	Canada	New Zealand	Australia
Export subsidies	x	—	—	—	—	—
Government trading institutions	—	—	x	—	—	—
Quasi-governmental entities	x	—	—	x	x	x
Export financing	x	x	x	—	—	—
Subsidies	(a)	x	x	—	—	x
Tax rebates and incentive schemes	x	—	—	—	—	x
Government agricultural programs	x	x	x	x	x	x
Price stabilization and support programs	x	x	x	x	—	—
Subsidized loans or loan guarantees credit assistance	x	x	x	—	—	—
Barter	x	—	—	x	x	x
Bilateral agreements	x	x	x	x	x	x
Government purchases:						
for aid	—	x	—	—	—	—
for social programs	—	x	—	x	—	—
for disaster aid	—	x	x	—	—	—
Conservation programs	x	x	—	—	—	—

(continues)

TABLE 6.2 (*continued*)

Type of Measure	European Comm.	United States	Japan	Canada	New Zealand	Australia
Adjustment assistance						
to workers	x	x	—	x	x	x
to industry	x	—	—	—	—	—
to communities	—	x	—	—	—	—
Subsidies to processors	x	—	—	—	—	x
Subsidized freight	(a)	—	x	—	x	x
Export licenses or permits	—	—	—	x	x	x

Source: Same as for Table 6.1. (a) = Some countries.

As a result of two years of research carried out at the University of Arizona, information on the regulatory and administrative practices affecting trade in meat is now available.[1] Most of the major measures that have been introduced by governments to regulate their meat production and trade are contained in Table 6.3.

TABLE 6.3 Technical Regulations Affecting Trade in Meat and Livestock

Nontariff Restrictions or Technical Regulations	Countries Imposing the Regulation	Purpose of Regulation
Cattle raising: Veterinary checks and health certificates	All countries	To ensure healthy animals and to prevent spread of disease and parasites

(*continues*)

TABLE 6.3 (*continued*)

Nontariff Restrictions or Technical Regulations	Countries Imposing the Regulation	Purpose of Regulation
Prohibition of the use of certain chemicals (nitrates), antibiotics, and hormone (growth stimulants such as diethyl stilbesterol)	All countries but not on all products	To reduce or eliminate possible adverse effects from the consumption of products treated or contaminated with chemicals, growth stimulants, antibiotics, drugs, or potentially harmful products.
Feed regulations	European Community Japan, Canada United States	To meet grading requirements
Distinction between grain fed (feedlot) cattle, range or pasture fed cattle and culled cows from dairy herds	United States Canada	
Government grades (cattle on the hoof)	United States	To achieve an efficient and fair cattle market
Certificates of origin (ranch or farms) and method of raising and finishing cattle	European Community, Japan	To ensure that grade standard requirements are met
Live cattle (for breeding or for feeder): Certificates issued by veterinary surgeon to accompany cattle	United States, Australia, New Zealand	To prevent the spread of disease

(*continues*)

94

TABLE 6.3 (*continued*)

Nontariff Restrictions or Technical Regulations	Countries Imposing the Regulation	Purpose of Regulation
Export permitted only from certified herds	United States, Australia, New Zealand	To maintain health standards and as a preventative measure
Transportation from cattle auctions: No wood (railings or stockade)	European Community	To prevent foreign matter entering products
Slaughter houses: Metal rails in holding pens and flues	European Community	To prevent splinters in hides or meat
Animals to be fed and watered if held longer than six hours	United States	Animal welfare
Areas where meat is handled must not have corners, areas must be painted specific color	European Community	Sanitation (technical standard) to ensure clean premises
Labor must shower and change before entering areas or facility designed for handling carcass and meat	European Community	Sanitation (technical standard) to prevent contamination of product
Labor must be healthy and have had a medical check	All countries	Health and sanitation standard to ensure wholesome product
Labor require periodic medical checks and to have a certificate of health before working	European Community	Health standard

TABLE 6.3 (*continued*)

Nontariff Restrictions or Technical Regulations	Countries Imposing the Regulation	Purpose of Regulation
No part of the equipment to be made of wood, *i.e.*, knife handles, pallets	European Community	To avoid splinters, foreign matter in the product (technical standard)
Veterinary check of animals ante-mortem	All countries	To ensure that health standard requirements are adhered to
Inspection for grading ante-mortem	United States	To meet grading requirements
Facility must be inspected and then issued with an export authorization certificate	European Community, United States	A technical standard designed to ensure high standards of hygiene
Public inspection of packing plants	European Community	As above
Killing floor: Knife to be washed in hot water after each slaughter	European Community	As above
Separate time for slaugher of suspect animals	United States, Canada, Australia, New Zealand, Japan	Health and technical standard to ensure wholesome product
Separate facility for slaughter of suspect animals	European Community	Technical standard
Butchering, removal of by-products and processing: Separate area (room) from the killing floor	European Community	Technical standard

(*continues*)

TABLE 6.3 (*continued*)

Nontariff Restrictions or Technical Regulations	Countries Imposing the Regulation	Purpose of Regulation
Carcass shrouding prohibited (the wrapping of carcasses in a muslin soaked in chlorinated solution)	Canada, European Community	Technical standard
Carcasses graded, stamped, and recorded; certificates to accompany meat shipment with relevant information required at port of entry (veterinary checks, grade, weight, etc.)	All countries	Technical standard; ensures that product is as consigned
Only specific inks or dyes can be used to stamp the carcass or meat	Some countries within the European Community	Technical standard prevents the contamination of the product with carcinogens
Carcass inspection by veterinary surgeon and specimens must be taken from meat, offal, and by-products for testing in laboratory	All countries	Health standard to ensure a wholesome product
Inspection of individual cuts	Japan	Ensures compliance with specifications
Meat "bone in" and "bone out" specification	All countries	Product specification
Meat cuts must be minimum size and weight	Most countries in European Community	Technical standard. Size of cuts to be large enough to reorganize the product and part of the animal.

(*continues*)

TABLE 6.3 (*continued*)

Nontariff Restrictions or Technical Regulations	Countries Imposing the Regulation	Purpose of Regulation
Packing: Shrink wrap carcasses not allowed	Greece	Technical standard
Labeling: Specific requirements such as authentication of grade specification, authority of signatures, origin of product, grade issued, health and sanitary certificates. Also, in many instances, the following are required: the nature of the product, the name and place of business of the manufacturer, packer and/or distributor, name of official distributor, name of official authority, and/or net weight of contents. Label placement size of type (print or lettering) and terminology used are sometimes specified by importing countries	All countries	Technical standard to ensure the product is legitimate and that the meat standards and specifications are met
Language specification for the information on labels and shipping documents. Marking specification on labels	European Community, Canada, United States, Australia, New Zealand	Technical standard to facilitate processing of documents
Meat processing: Cooked meat requirements including temperature and humidity control and length of cooking time	All countries	Health standard to eliminate salmonella and ensure destruction of other bovine diseases

(*continues*)

TABLE 6.3 (*continued*)

Nontariff Restrictions or Technical Regulations	Countries Imposing the Regulation	Purpose of Regulation
Data on date of production and shelf life of canned meat	United States, European Community	Health standard to ensure wholesome product
Standard sizes of containers or packages	Canada	
Storage constraints: Distinction in quota system between frozen and fresh-chilled meat	United States, Japan, Canada European Community	Technical standard to differentiate the meat producer

Source: Interviews in 1982-84 by Mark B. Lynham with individuals in the meat industry in the United States, the National Cattlemen's Association, United States Meat Exporters Federation; Meat and Livestock Commission in the United Kingdom; Australian Meat and Livestock Corporation, New Zealand Meat Producers Board, United States Department of Agriculture, agricultural and veterinary attachés in the Australian, New Zealand, United Kingdom and United States Embassies.

Quotas on the imports of red meats have been in operation in the United States since the 1960s; the President has the power to adjust these quotas as seems appropriate in the circumstances.[2] Various amendments have been made to the earlier legislation on quotas the latest being the Meat Import Act of 1979 (Appendix D). This law supersedes the Trade Agreement Act which was approved earlier in the same year. The Meat Import Act in fact increased the quota provided by the Trade Agreement Act and, in practice, this legislation has not so far proved to be a major trade restraint. It is important, however, in that it lays down a formula for determining the annual quota on beef imports into the United States.

In addition to the quotas on red meat, the United States maintains a standby quota system for the importation of live animals at the Mexican and Canadian borders. More importantly, however, Section 22 of the Agricultural Adjustment Act provides for the imposition of quotas if imports interfere, or are likely to interfere, with government price support and stabilization programs.

Laws on health and sanitary requirements, together with product standards and technical requirements, are currently unpopular with exporters to the United States. This unpopularity derives not from any difficulty in meeting standards but primarily because of the work involved in complying with the administrative detail.

Controls in the European Community are epitomized by the Third Country Meat Directive (72/462/EEC) which has been negotiated and renegotiated by the member countries for well over a decade. It is concerned with the health and veterinary inspection of bovine animals, swine and fresh meat imported into the Community; it also covers technical standards for the processing, handling and shipping of red meat destined for the Community. The original version of the Directive was issued in 1972 but not all Community countries followed the rules and regulations with respect to the import of meat and livestock. Some countries—for example, the United Kingdom—were more lenient. The latest version of the Directive (1985) took effect at the end of a grace period (January 1, 1987 for the United States) and not enough time has yet elapsed to assess its impact. It has been feared that it will greatly reduce, if not eliminate, meat imports from many countries that find it hard or costly to comply with the standards required. Of course the major exporters of beef, such as Australia and New Zealand, are already adapting their industries to the Directive in order to maintain their meat exports to the European Community. In the United States, however, meat processors may cease shipping meat to Europe rather than make the large capital expenditure which would be needed in order to conform to the requirements of the Directive.

In Japan, the authority for the imposition of agricultural quotas is embodied in the 1961 Price Stabilization Law for Livestock Products which gives the Ministry of Agriculture, Forestry and Fisheries (MAFF) power to set import quotas with a view to supporting domestic meat prices.

In addition, health and sanitary regulations for all products are applied with rigor and standards for animal and animal product imports are no exception. Stringency in itself should not be objectionable, so long as the rationale for it is scientifically sound and the enforcement non-discriminatory. In general, Japanese procedures for inspection and standards of contamination control cannot justly be labeled barriers to trade. Instances can be found in which particular rules for entry into Japan or the official interpretation of test results in Japan are at variance with those in other countries. But discrepancies are bound to exist and until all other countries reach concensus, the peculiarities of Japanese regulations should not be singled out.

Legislation in other countries is of a similar kind to that already briefly outlined. Next to the direct quota, the most important legislation affecting meat and livestock trade (and in fact, agricultural products in general) refers to health and sanitary regulations; these will now be considered in detail.

Health and Sanitary Regulations

It is safe to say that every country that engages in the production or trade of livestock and meat imposes restraints in the form of health and sanitary regulations. One major class of nontariff barrier in the meat and livestock industry derives from legislation and regulations relating to the control of disease and parasitic problems in cattle the spread of which can in some cases be harmful to consumers and costly to producers. Consumers expect their governments to ensure that meat sold by suppliers is safe to eat, while producers rely on governments to prevent infection and to minimize the spread of parasites and disease, especially contagious diseases such as foot-and-mouth disease or *aftosa*. A study by James Simpson and Donald Farris of the University of Iowa carried out in the early 1980s contains a detailed taxonomy of these diseases.[3]

Australia and New Zealand enforce strict health standards and consider themselves to be free of cattle diseases. Japan, Canada and the United States have eradicated or controlled many cattle diseases and they now protect domestic herds by closing their borders to imports of fresh meat and live animals from most of the rest of the world. The incidence of disease varies in Europe. Outbreaks occur occasionally, especially foot-and-mouth disease, chiefly as a result of inadequate inspection facilities or because the inspection requirements have been relaxed. Most European countries innoculate against foot-and-mouth disease, but the United Kingdom destroys infected herds.

Access to the British market is apparently easier than to the American; some countries that are excluded from the United States by prior embargo may export to the UK provided that actual shipments pass customs inspection. But the more liberal import procedures are not without disadvantages for exporters. When, for instance, a Uruguayan packer ships beef to the United Kingdom he takes a risk that either his consignment or any other from his country will not pass the tests; this risk can be costly. In 1969 Uruguay was contaminated with foot-and-mouth disease and the disease was transported in the bone of red meat products imported by the UK. Upon verification of the disease in June 1969, British officials revoked Uruguay's right to export beef products; for Uruguay there was a loss of trade of almost 33,500 tons in the following year.

The meat and livestock industry is also subject to regulations covering the feed, chemicals, and antibiotics administered to the animal from birth to slaughter. The processor of the slaughtered steer, cow, or calf must comply not only with the regulations in his own country but also with the regulations of the countries to which the product is being exported. Regulations and standards concerned with food additives are especially susceptible to change and the changes can be costly. It is often long after the industry has incorporated a certain chemical treatment or feed additive into production or marketing techniques that evidence emerges of deleterious effects on public health. In such cases it is difficult enough to strike a fair balance between the risk to the public and the loss of an investment not only to domestic producers but also to foreign suppliers. When Canada banned the use of the hormone growth stimulant, diethyl stilbesterol, and later banned the hyper-cholorination wash used in the treatment of beef carcasses, exports of beef from the United States to Canada ceased altogether.

From calving to supermarket freezer the government interposes a succession of standards, regulations and requirements for a wholesome product in the beef market. Undoubtedly, government interference is warranted and most of the measures spring from legitimate precautionary motives. As the Food and Agriculture Organization (FAO) has observed correctly: "Although restrictions based on sanitary grounds are impediments to trade, it has to be stated that they are to a large extent justified."[4] What should be of concern are the "unjustified" restrictions. Among these are the following: (i) inconsistency in the application of standards which lead not only to arbitrary decisions but also to discrimination and (ii) the imposition of safeguards or remedies where the cost is disproportionate to the risks involved.

An example of the first can be found in the European Community before the 1985 amendments to the Third Country Meat Directive became operative. The slaughtering, storage and transport employed in the processing and movement of meat within the Community had to conform to the specific health and sanitary standards of each member government. A paramount problem, therefore, was the need for a common minimum standard code. Attempts were made in the early 1970s to establish such a sanitary code but acceptance was blocked by strong opposition on two fronts. Some member governments objected to liberalizing their standards on specific health regulations while other governments were hesitant to allocate additional resources in capital and manpower to toughen regulations and upgrade inspection procedures to comply with a uniform health code. The resulting discrepancies placed a serious economic burden on exporting countries.

The divergence of these standards was not only wide but sometimes whimsical. Some countries required that carcass meat be delivered with all the lymph nodes opened by the veterinary service of the exporting country as proof of inspection. Other countries required that at least one lymph node be left intact so that their own veterinary service could carry out an inspection. Provisions in the United Kingdom allowed meat to be imported without any lymph glands at all so long as an official certificate from the exporting verterinary service verified that the meat was healthy.[5]

During the 1970s there were also variations in packaging requirements. In France, frozen beef weighing from 20–30 kg had to be packaged with inner wrappings of polyethylene and outer wrappings of jute or blue and white plastic. Belgium and Luxembourg required similar packaging material but weight classifications for approved shipments were from 40–50 kg for jute packs and from 30–35 kg for cartons. Packing and marking practices required by Italy were based on the Italo-Argentine Meat Agreement of September 1970 and these differed from those of all Italy's neighbors. Shipments of boneless beef had to have an inner packing made from transparent, colorless and sterile material (such as polyethylene or cryovac film), with an outer pack of cardboard or wood.[6] The Italian requirements were not favorably received by international traders, who preferred the cheaper and traditional jute packing.

Similarly, transport regulations were diverse throughout the Community. Some countries required that temperatures should be −18°F throughout a sea voyage although containerized ships routinely maintain a temperature of −10°F.[7] Requirements in some countries were that inland deliveries of meats in containers must be refrigerated, even though it had been proved that unrefrigerated containers maintain satisfactory temperatures up to 40 hours. In addition, veterinary procedures of some member countries required that refrigerated vans be opened at the frontiers, thereby increasing the risks of contamination and spoilage. Attempts at harmonization within the Community met with varying degrees of acceptance.

A further example of an "unjustified" restriction, may be found in the administration of the live cattle quarantine in Japan. Were the tariff quota the sole deterrent to importation of feeder and slaughter cattle, it is likely that imports would greatly exceed current levels. In the late 1970s, live cattle imports rose quickly from a few thousand to around 15,000 in 1979, a peak which they have not surpassed since. The availability of quarantine facilities in the country has placed a strict limit upon further import growth. The facilities currently available to house animals for the mandatory minimum period—five days in the case of slaughter cattle and fifteen days in the case of feeder stock—cannot accommodate larger numbers. Yet the government will not construct additional capacity, nor will it agree to permit quarantine at sea aboard barges specially fitted out for this purpose.[8] The reluctance of Japanese officials is understandable. Beef exporters could evade the import quota by shipping live animals instead. Feedlot operators in Japan are their eager accomplices. In order to constrict this loophole and reinforce the quota system, the government leaves a barrier in place here and it is effectively insurmountable.

Other Specific Nontariff Measures

In addition to health and safety regulations, it will be seen from Table 6.1 that a variety of nontariff measures are used to control meat imports in addition to health and sanitary regulations. Some of the measures shown in the table are now examined in more detail.

Grading Differentiation: An obvious example of this practice is the beef grading system adopted in the United States. This grading system affects both imports of beef into the United States (and therefore producers and exporters in other countries) and producers and exporters in the United States. On the import side, the specifications laid down by the United States Department of Agriculture (USDA) in essence preclude the importation of beef from regions producing range-fed beef.[9] Australian and New Zealand beef comes mostly from range-fed cattle, while the Canadian beef is somewhat similar to that produced in the United States. Because the Canadian beef grading descriptions and nomenclature are different from those of the United States, however, beef has to be regraded for import into the United States which means that extra costs are incurred. To avoid these costs, most fresh, chilled and frozen Canadian beef enters the United States classified as lower quality beef.

The United States is not the only country to use the grading system to control imports of beef. In fact, negotiations on beef grading were conducted in the GATT during the Tokyo Round of multilateral trade negotiations and a definition of prime and choice beef (recommended by the USDA) was accepted by the contracting parties to the GATT. Additional requirements were included in the GATT specification for high quality beef, however, in that special requirements were laid down in relation to the age of the steers and the kind of feed they received in the last three months or so before they were shipped for slaughter.

The high quality beef description laid down by the GATT reflects the preferences of producers and traders in the United States but it is less well suited to consumer tastes. Japan and the United States are the two major importing countries whose consumers prefer more heavily marbled meat, in spite of the trend towards leaner beef for economic and dietary reasons.[10] But the Meat Export Federation in the United States has reported that there is still a substantial demand for leaner beef in the countries of the Community and very little demand for high quality beef (which is heavily marbled with fat) as currently defined by GATT.

On the other hand, under the current grading regulations, leaner beef imported into the Community is subject to a variable levy whereas a levy-free quota of 10,000 metric tons has been negotiated for high quality beef as defined by the GATT. Importers in the Community would like to import leaner beef under the high quality beef quota although the suppliers in the United States would have some difficulty in meeting the required standard.

In 1981, a proposal was submitted to change the official USDA standards for grades of carcass beef. The initial impetus for the change came from the National Cattlemen's Association, which advocated less marbling in all grades. Specifying a leaner product for all grades would reduce production costs and improve net returns for the cattlemen. In response to strong lobbying against the initiative by retail butchers and the hotel-restaurant trade, it was decided not to adopt the proposal.[11] The Meat Export Federation, in commenting on the proposal, indicated that a "lowering" of the beef grading standards—meaning a reduction in the fat content—would improve the chances of expanding the beef export trade to the European Community. It is clear, therefore, that in this particular case, the grading specifications are acting as a nontariff trade barrier to the export of beef from the United States to the Community.

Licensing Arrangements: Mention already has been made of the system of quotas which most countries use to control the import of meat and livestock as well as other products. Such quota systems usually go hand-in-hand with a licensing system as this will enable the quotas to be monitored. These licensing procedures, however, do present a number of obstacles to trade but not least in that the procedure for obtaining import licenses is often complex and the information which has to be supplied to the licensing authorities is considerable. This is especially irksome when licenses are issued for short periods of time as was the case with import licenses for high quality beef in the European Community: most import licenses   in the

Community are issued every three months but high quality beef permits have to be renewed monthly.

Administration of Restrictions: The way in which barriers to trade are administered can in themselves add to trade distortion. For example, in the meat trade the Ministry of Agriculture in Japan manipulates the import quota for beef in order to keep domestic prices within a limited range of variation. Control over trade in pork is sustained through the use of a variable levy where the objective is also to maintain stable domestic prices. Through the use of the variable levy, the physical quantity of pork imports is regulated in accordance with price control policies. The quota and variable levy exert the principal restrictive influences on the meat trade in Japan but the manner in which the Japanese government administers them tends to compound the distortion and the impediment to trade.

Another aspect is that would-be exporters to Japan face discrimination, unnecessarily cumbersome regulations and inconsistencies in administrative procedure. Moreover, quotas are not always global but may routinely favor some areas or countries at the expense of others. This is done by specifying certain requirements which can be met only by some countries. For example, until recently part of the general quota in Japan designated for high quality beef was defined in accordance with grain feeding standards and grades in the United States.

It is not unlikely that the American share of the Japanese market would be small were it not for the preferential access which the high quality beef specifications afford. Because the degree of fat marbling is low by Japanese standards, American choice ("high quality") beef does not command a high price in the Japanese market. Although frozen grain-fed beef from the United States incurs higher transport costs than Australian grass-fed beef, its wholesale price is significantly lower than Australian chilled beef and only a little more than Australian frozen beef. Bilateral agreements have overruled a competitive disadvantage and Japan now imports increasing amounts of beef from the United States at the expense of trade with Australia.

In the mid–1980s, however, Australian officials successfully negotiated a change in the definition of high quality beef acceptable to Japan. Although Australian exports of high quality beef to Japan are still relatively insignificant this adjustment in beef specification may open up a large new market for them. This illustrates very clearly the significance of administrative detail in international agricultural trade.

Administrative discrepancies can also occur in the arrangements which are made with exporters. For example, would-be exporters of beef to Japan have to tender their bids to a local subsidiary of a Japanese trading company. Not only is this procedure indirect and time-consuming, it often introduces discrepancies in the contract for which bids are made. Local offices may, and do, report product specifi- cations cations that differ from those originally issued in Japan. As a result, bids from, say, an American and an Australian firm are not always comparable.[12]

Administrative Guidance: Another nontariff measure that deserves mention is the not uncommon practice of "administrative guidance." Here again, the methods used in Japan can be cited as an example. In some cases, administrative guidance is built into standard import pro- cedures. In other cases, it is an extraordinary resort which is employed when regular means of control are inadequate. It should be understood that administrative guidance is in all cases void of legal power. In- tended as advice rather than coercion, it is nontheless a powerful form of control over the market and over trade. For example, the Japanese Ministry of Finance has responsibility for determining the annual quota of duty-free live cattle. In 1982, a quota of 11,000 head of feeder calves was recommended by the Ministry of Finance. The Ministry of Agricul- ture, however, was concerned about the current low price of domestic feeder calves and it prevailed upon the four livestock cooperatives who share the quota to limit their imports to just under 4,800 head that year. Similarly, over-production of pork caused domestic prices to fall sharply toward the end of 1979. After pressure from domestic pork producers, Japanese importers and processors curtailed their order for imported pork during the early months of 1980. The influence of the Ministry of Agriculture was suspected but the Ministry contended that the action was "voluntary." These claims, of course, need not be incon- sistent and both may well be true.[13]

Evasion of Nontariff Measures

Where nontariff measures become highly distorting or discrimina- tory, importers and exporters will try to evade them. Restrictions may, of course, be imposed on a complementary product to prevent evasion. An example of this is the tariff quota which is applied to live cattle imports into Japan. A certain number of feeder cattle are imported free of duty but any above this number pay a high tariff rate. Japanese im- porters and processors would prefer to import more live cattle from Australia because they are relatively inexpensive and the cost of beef production could be reduced. But the tariff quota on live cattle is spec- ifically designed to circumvention of the quota on meat imports.

Evasion of the quota has, however, been taking place by different means. Since World War II a 25 percent *ad valorem* tariff was levied on beef offal as well as on other cuts of beef. The government of Japan removed the tariff from offal completely in 1982. While the Japanese definition of edible offal includes hearts, livers, tongues, tripe and other viscera, most of the meat traded under this category free of quota restriction has consisted of lower grade stewing beef. In grain-fed beef from the United States (but not range-fed from Australia) these parts are succulent enough for use as table meat. Firms on both sides of the Pacific have taken advantage of this loophole in the import regime. The growth of trade in offal has been rapid and the United States now supplies Japan with more offal than high quality beef. By 1982, Japan was importing 32,000 tons of high quality beef cuts and 44,000 tons of so-called offals.

More in a spirit of accommodation rather than evasion, one American meat packer has vertically integrated its operation in order to reduce the problems involved in obtaining an export processing license from the European Community and also to minimize the uncertainty in the beef and livestock trade. It has set up an import and wholesale distribution subsidiary in the United Kingdom and has integrated its operations in the United States including shipping facilities. The arm of the firm located in the United States fattens and finishes cattle, processes and packs the meat and ships directly to its subsidiary in the United Kingdom from which the order will originate. The United Kingdom subsidiary promotes the product, emphasizing the differential qualities and using in-house quality brand names and descriptions. The required papers and certificates are easily obtainable. Consignments can be taken as single units for quality, health, and technical standard requirements. Economies of scale can be fully utilized to supply the desired product—such as leaner grain-fed beef—at a reasonable and competitive price, even with the variable levy and the 20 percent import duty charged against non-quota beef. The venture is fairly new but is proving profitable; all existing health and sanitary requirements have been met.

Another example is a group of Japanese investors who recently purchased a large ranch in California. Their intention is to raise grain-fed steers, taking advantage of the efficient American system; the live cattle will then be shipped to Japan for finishing and slaughtering. This strategem avoids the strict grade and health certification that is required by Japan on all beef imports, which can add up to $0.40 per pound to the price of the product. The major consideration is that live cattle and feeders for slaughter are not considered under the meat import quota system but under a tariff quota system.[14] The limited quarantine facilities in Japan, however, still present a barrier to free trade.

Some of the procedures frequently used to evade restrictive trade practices are more subtle and at times even covert. Bribery has already been mentioned as a means of circumventing restrictions. Personal contact represents a more conventional method. Exporter and importer try to ensure that they have a good personal relationship (both social and business) with government officials and administrators. Many a problem at the port of entry is resolved over a meal or a drink.

Another fairly common practice in the European Community is the shipping of consignments by third countries to less highly restricted nations within the Community who then re-export—legally or illegally —to the ultimate designation within the Community; this allows advantage to be taken of the more tolerant trade regulations among Community members. The nation serving as middleman invariably finds such transactions to be extremely profitable because of price differences within the Community itself.

The number and the proliferation of nontariff trade barriers are limited only by the ingenuity of the importing countries. At the same time, if the price is right and if the opportunity permits high returns, firms and individuals will seek ways to circumvent trade barriers. In the European Community and in Japan, the beef trade presents opportunities for high returns, where the international price of certain categories of beef is comparatively low and the domestic retail price relatively high. The high domestic price could be due to three possible factors either singly or in combination—restrictive import practices, domestic trading structure, and/or farm support policies. In both the European Community and Japan, the consumer is penalized through having to pay higher prices than necessary for beef. The very strict health, hygiene, and technical standards that are required in the beef and livestock product industry by governments induce individuals, firms and even government agencies to flout the system in search of high returns.

Notes and Sources

1. Mark B. Lynham, *Nontariff Agricultural Trade Barriers: Livestock and Meat Legislative Regulation Devices as they Affect International Trade Between Industrially Developed Countries.* Unpublished Master's Thesis, Department of Agricultural Economics, University of Arizona, 1983.

2. The most sweeping action taken under these executive powers was the complete relaxation of quotas in June 1972 because of inflationary pressures on domestic prices of red meat. This action was superseded by the Meat Import Act of 1979 which relaxed the quota restrictions.

3.   James R. Simpson and Donald E. Farris.   *The World's Beef Business*, (Ames, Iowa: Iowa State University Press, 1982).

4.   *Nontariff Barriers to International Meat Trade Arising from Health Requirements*, (Rome: Food and Agriculture Organization, 1973), p. 1.

5.   *The Market for Manufacturing Grade Beef in the United Kingdom and the European Economic Community*, (Geneva: UNCTAD, Secretariat, 1971), p. 53.

6.   *Ibid. , pp.* 76, 85, 99 *and* 127.

7.   Seminar held by the New Zealand Meat Producers Board in London, 1972.

8.   *Japan: Background on Agricultural Trade Barriers*, mimeograph,   Foreign Agricultural Service, USDA (Washington: U.S. Government Printing Service, undated), pp. 5-6.

9.   Range-fed cattle are leaner than grain-fed cattle and thus fall into a lower quality category in the United States.

10.   *United States Federal Register*, Washington, Vol. 46, No. 250, December 30, 1981, p. 93054.   Japan's high quality beef is even more heavily marbled with fat than choice grades in the United States.

11.   *Ibid.*

12.   *Agenda for Action*, Report of the Japan-United States Businessmen's Conference, Washington, July 1983, p. 88.

13.   *Japan: Background on Agricultural Trade Barriers, op. cit.*

14.   The quotas are set periodically, generally about twice a year.   If the import exceeds quota levels, then a more restrictive tariff is applied. See William T. Coyle, *Japan's Feed-Livestock Economy*, Economic Report No. 177, Foreign Agricultural Service, USDA, 1983.

7

Technical Barriers to Trade in Poultry[1]

In this chapter a case study is presented to show how health and sanitary regulations may be used as a protective trade measure. In the early 1980s the governments of the United Kingdom and France engaged in a lengthy skirmish to protect their poultry farmers from cross-Channel imports. A single event in a series of trade retaliations that took place during that period will be described in this chapter. The comparative cost structures in poultry farming in the two countries will also be examined with a view to assessing the trade distortions caused by technical and other barriers to trade in poultry.

"Chicken War" 1978–82

The opening round in what has come to be popularly known as the European version of the "chicken war" was fired by the French. French poultry farmers in the late 1970s were able to obtain various forms of economic assistance—mainly subsidies—from the government to establish several large, efficient packing houses on the Brittany peninsula. Poultry farmers in the United Kingdom became aware of their growing inability to compete with the French.

On August 27, 1981, the Minister of Agriculture in the United Kingdom announced a precautionary measure to prevent the spread of Newcastle disease in the United Kingdom:

Beginning September 1, a compulsory slaughter policy is being reintroduced to deal with any future outbreaks of Newcastle disease ("fowlpest") and the use of vaccine will be prohibited.

111

At the same time, imports of poultry meat and eggs will be permitted only from countries which are free from Newcastle disease, which ban the use of vaccine and which also apply compulsory slaughter in the event of an outbreak of the disease.[2]

The ban had little to do with disease infestation. In fact, the United Kingdom had been free from Newcastle disease for five years prior to the announcement. Of greater significance was the domestic political activity which inspired the ban. The announcement followed on the heels of a vigorous campaign against poultry imports by the National Farmers' Union (NFU) in the United Kingdom. The NFU had campaigned primarily against imports of French turkeys, which it alleged, were sold in the United Kingdom at nine pence below production costs thanks to subsidies given by the French government to French poultry farmers. (Ironically, turkeys are not susceptible to Newcastle disease.)

The restriction on poultry imports into the UK eliminated supplies from all countries except Ireland, Denmark and Sweden which could comply with the new trade requirements. The United Kingdom continued to receive over 90 percent of its poultry imports from member states in the Community, but the pattern of trade shifted noticeably. Historically, Danish exporters traded with Germany and to a lesser extent with the United Kingdom. During the year of the ban on poultry imports, however, Danish exports to the United Kingdom increased by 752 percent to 21,300 tons in 1982 from a 1975–1981 average of 2,500 tons. Ireland's exports to the United Kingdom increased by 20 percent to 6,600 metric tons in 1982 from a pre-ban average of 5,500 tons.

Dutch exports to the United Kingdom were an inadvertent casualty of the ban. Exports from the Netherlands averaged 13,200 tons from 1975–1981, but in 1982 only 1,300 tons of Dutch poultry were permitted to enter the United Kingdom. The balance of normal UK-Netherlands trade was absorbed by Germany. The German share of Dutch exports increased by 12 points from its 1981 level of 66.4 percent to 78.4 percent or an additional 6,000 tons.

It is ironic that the nation to which the ban was directed seems to have been least affected by it. And yet, while the impact on France is not so obvious as the impact on Denmark and the Netherlands, the losses were nonetheless substantial. France's exports to the United Kingdom dropped just 1,670 tons from its 1975–81 average of 2,370 tons to 700 tons in 1982. In the absence of the ban, however, trade volumes could have been significantly greater than the French historical average. (French exports to the world grew at an unprecedented average annual rate of 23 percent during 1975–82 from 82.7 to 344.9 thousand tons.) Contracts for 70 tons a week of turkey meat, 3,000 tons of whole turkeys and 100 million eggs were nullified by the ban. An estimated 2,000 tons of frozen broilers were also denied entry. In total, French poultry meat and eggs to the value of $22.2 million were banned from the UK market.

Community Response to the "Newcastle" Ban

The United Kingdom was within its rights, since the European Community allows member states to institute health regulations at stricter levels than called for in the common agricultural policy. Nor was formal complaint of discriminatory action warranted because the restriction applied to all nations and was supposedly used for health considerations, not domestic market protection.

Nevertheless, the European Commission argued that the ban was unjustified by the "extent of its coverage and the absolute nature of its effect."[3] (Denmark operates a similar ban, but its controls are more flexible as it maintains a vaccine check at the border.) Secondly, because British producers were allowed to slaughter and market their own vaccinated birds, the action was clearly discriminatory. Thirdly, the ban had not been phased so as to provide exporters a reasonable adjustment period as had been allowed to the producers in the United Kingdom; it was announced and imposed four days later. (Interestingly enough, the United Kingdom itself was still marketing vaccinated poultry in 1982.)

The French poultry exporters moved quickly in order to comply with the new regulations but the regulations continued to be used by the United Kingdom to restrict imports; the case was therefore taken to the European Court of Justice early in 1982.

The Court ruled that the ban was obviously not part of a seriously considered health policy but a "thinly disguised import restriction whose real aim was to block, for commercial reasons. . . , imports of poultry and poultry products from other member states, particularly France."[4] Had the United Kingdom been acting out of true concern for animal health, it was reasoned by the Court, it would have reopened its market to French produce after France had fulfilled the three conditions laid down by the United Kingdom in August, 1981.

It was also noted that the ban did not apply originally to exotic birds, which pose a greater threat of infestation than vaccinated poultry. Furthermore, it was argued that outbreaks of the disease had fallen dramatically in all member states, even in some countries where not all birds had been vaccinated. Thus, the action by the United Kingdom was judged to be a violation of the legal principle of proportionality, since the damage against trade exceeded any potential benefits to animal health in that country.

The government of the United Kingdom, however, continued to use a series of negotiating tactics to secure their borders long enough for the 1982 Christmas trade to be reserved exclusively for their domestic industry. In total, they were able to get sixteen months of protection, including two Christmas seasons, out of the poultry import ban.

This particular case (and others concerned with agricultural products) raises doubts about the benefits of Community membership. Poultry farmers in the United Kingdom complained that because of subsidies given to poultry farmers in France by the French government, poultry from France was more competitive. If this is true, what shortcomings in the common agricultural policy allow the French government to make such outlays and are they conducive to a fair and competitive environment?

Production subsidies are said to be a violation of the spirit of the Treaty, yet they are permitted by the Articles in the Treaty of Rome referring to agriculture. Tariffs, on the other hand, are explicitly prohibited. These are interesting provisions, given that subsidies can alter the competitive balance as effectively as import tariffs. Since the European Community provides no recourse against member states which indirectly subsidize production, individual member states often stand to gain from production subsidies. This suggests that unless production subsidies are also made illegal, further trade wars are likely in the future and the use of nontariff measures will increase.

Comparative Costs in Britain and France

Although poultry farmers in the United Kingdom complained of unfair competition because of the subsidy given to French poultry farmers, it would be useful to examine the variation in costs between the poultry farmers in the two countries. Private and social profitability between the poultry sectors of the two countries is used as a proxy to determine relative competitive advantage and the effects of government policy upon the competitive balance. Costs of production tables have been constructed which separate private and social costs. Private costs are those costs which the producer faces. This includes the price of factors of production together with intermediate inputs plus any taxes incurred by the producer. Social costs include private costs plus capital grants, interest rate subsidies and rebates on value added tax on buildings and equipment, but not on variable costs. The effects of government policy on production are derived by subtracting private costs from social costs. A net positive value denotes government assistance. A negative balance indicates that taxes exceed subsidies. A value of zero suggests that government policies cancel one another out or that policies simply do not exist as in the case of the United Kingdom.

The effects of taxes and subsidies on French production costs are shown in Table 7.1. Costs are given in constant ($1980) prices and rebates on value added taxes are discounted at the annual interest rate to account for the time value of withheld tax contributions. (In general, value added taxes are paid at the time of purchase and then refunded at the end of the production cycle or calendar year.)

TABLE 7.1 Social and Private Costs of Poultry Farming, France, 1978-82 ($US 1980[a])

	1978	1979	1980	1981	1982
Fixed Cost	.093	.093	.089	.080	.100
Less:					
Capital grants or rebates	.032	.027	.038	.036	.036
Fixed cost (net)	.061	.066	.051	.041	.064
Variable costs	1.088	1.064	1.005	1.002	.939
Less:					
Rebate of value added tax	0.67	.065	.061	.059	.048
Training grant	.005	.004	.005	.004	.004
Net variable costs	1.016	.995	.939	.939	.887
Social costs	1.109	1.088	1.028	1.019	.987
Private costs	1.1077	1.061	.990	.983	.951
Policy effect	.032	.027	.038	.034	.036

Sources: *Eurostat*: *Agricultural Price Statistics*, 1971-1982 (Luxembourg: Statistical Publications Office of the European Communities, 1983); *Les coutes de production de poulet de chair* (Paris: Institut Technique de l'Agriculture, 1978–82).

(a) Cost per kilogram, live weight, less than 16 weeks old.

The complicated system of taxes and subsidies in France appears to give with one hand and take with the other. Subsidies are offset by taxes but are then partially restored by rebates. Contributions from the French government include subsidized interest rates, grants for buildings, equipment and labor, and rebates on value added taxes on capital expenditure. French producers received an interest rate credit amounting to 5 percent on the purchase of new buildings and equipment; a subsidy of 0.3 to 1.22 cents per kilogram of live weight poultry in production costs. French producers were also assessed, however, with a value added tax of 7 percent on capital expenditure, but the full amount of the tax is refunded. French poultry farmers incurred only the opportunity costs of the tax for the time it was held by the government.

On the other hand, they received a direct subsidy, amounting to 27 percent of total wages paid, to defray the expense of training new workers. Production costs were reduced by 0.5 cents per kilogram live weight by this grant. But part of this saving was offset by payroll taxes, a tax on vocational training and apprenticeships and a tax on meat inspection. These taxes added from 0.2–0.4 cents per kilogram live weight to production costs. Value added taxes on the purchase of chicks, feed and veterinary services are also assessed against French poultry farmers but they are refunded upon application to tax officials. Total grants ranged from 1.5–2.3 cents per kilogram live weight during the five year period examined. Value added taxes cost producers from 0.1–0.2 cents per kilogram live weight in opportunity costs. The net effect of French government policy on poultry production was a subsidy of 2.7–3.8 cents per kilogram live weight.

Estimates of private and social costs of production in the period 1978–82 in the United Kingdom are given in Table 7.2. Costs are deflated and converted from national currency on a per bird basis (as published by the National Farmers' Union) to United States dollars per kilogram to facilitate comparison with costs in the French poultry industry. Fixed costs are not published for the United Kingdom and they were therefore estimated. The variable cost figures are taken from the quarterly averages published by the National Farmers' Union. No government transfers or financial assistance is provided to the British industry and social costs therefore differ from private costs by the amount of value added tax on buildings and equipment.

The two tables show clearly that the subsidies given to French poultry farmers provide an economic incentive for technical innovation because they receive both a rebate on the value added tax paid on buildings and equipment and an interest rate subsidy. The poultry farmer in the United Kingdom, on the other hand, has to pay the market rate of

TABLE 7.2   Social and Private Costs of Poultry Farming,
United Kingdom, 1978-82   ($US 1980[a])

	1978	1979	1980	1981	1982
Fixed costs	.126	.127	.123	.126	.127
Variable costs	1.180	1.123	1.009	.976	.945
Social costs[b]	1.305	1.248	1.130	1.101	1.070
Private costs	1.306	1.250	1.132	1.103	1.072
Policy effect	−0.001	−0.002	−0.002	−0.002	−0.002

Sources: *Eurostat*: *Agricultural Price Statistics*, 1971–82 (Luxembourg: Statistical Publications Office of the European Communities, 1983); *Quarterly Broiler Bulletin*, London, National Farmers' Union, various issues, 1978-82.
   (a)   Cost per kilogram, live weight, less than 16 weeks old.
   (b)   Private costs less value added tax on buildings and equipment.

interest on all borrowed funds, as well as a value added tax of 15 percent on the full purchase price of buildings and equipment. This tax adds 0.1 cents per kilogram live weight to production costs and it can only discourage investment in new plant and equipment by poultry farmers in the United Kingdom.

Impact of Subsidies and Taxes on Prices

   Social and private production costs for each country are compared in Table 7.3 and this shows that costs in France are lower than in the United Kingdom.   Price differences that result from economic efficiencies and relative levels of government taxation are revealed by comparing the social costs of each industry.   French producers had a significant social cost advantage throughout the five-year period;   their

TABLE 7.3   Comparison of Social and Private Costs of Poultry
Farming, United Kingdom and France, 1978–82
($US 1980[a])

	1978	1979	1980	1981	1982
Social Costs:					
United Kingdom	1.305	1.248	1.130	1.101	1.070
France	1.109	1.088	1.028	1.019	.987
Cost difference	.196	.160	.102	.082	.083
Private Costs:					
United Kingdom	1.306	1.250	1.132	1.103	1.072
France	1.077	1.061	.990	.983	.951
Cost Difference	.229	.189	.142	.120	.121

Source: Tables 7.1 and 7.2.
   (a) Cost per kilogram, live weight, less than 16 months old.

social costs were less by a range of $83 to $196 per metric ton of live
weight.
    The level of government assistance is shown by the difference be-
tween private and social costs.   Private costs in the United Kingdom
exceeded social costs by $2 per metric ton of live weight. No govern-
ment assistance is provided to the poultry industry in the United Kingdom but in France, government assistance ranged from $27 to $38 per
metric ton of live weight. French private costs, however, ranged from
$119 to $229 per metric ton less than the costs of producers in the
United Kingdom over the five-year period and this indicates that the
major portion of the difference in cost between the two countries is explained by differences in efficiency rather than differences in government
subsidies.

Efficiency Differences

The French broiler chicken weighs approximately 60 grams less, uses ten percent less feed and requires only 44 feeding days as compared to 53 in the United Kingdom. Nine fewer days on feed and an average difference of one week between rotations (14 days for France versus 21 for Britain), translates into an average of 6.8 rotations per year, or 27 percent more than the 4.86 rotation in the United Kingdom. Furthermore, more efficient use of poultry housing contributes significantly to the French comparative advantage. French broiler statistics report an average stocking density of 22 birds per square meter for the years 1978-82, while stocking densities in the United Kingdom averaged 18 birds per square meter. Given the rotation rates and an average mortality rate of four percent for both countries, French production is 54 percent more efficient per unit of housing areas (130 as compared with 84 broilers per square meter per year). Average fixed costs are 21 cents per kilogram live weight lower when using the French production methods than if the United Kingdom methods are used.

France could have participated in the United Kingdom poultry market on a greater scale much sooner if sales to other member states had not incurred an additional levy as a result of the Community's common agricultural policy. Monetary compensatory amounts (MCAs) were introduced in 1969 as a stop-gap measure after the devaluation of the French franc and the revaluation of the German mark threatened to undermine the common pricing system. The Community desired to isolate the agricultural sector from the inconvenience of currency adjustment and reintroduced an intratrade levy to stabilize national prices.

The MCA is a tax on French exports and a refund for French imports. Monetary compensatory amounts apply to products which are subject to intervention and to products whose prices depend upon the price of products subject to intervention (for example, poultry). Monetary compensatory amounts assessed against French poultry exports were equal to $16.00 per metric ton live weight in 1978 and declined annually to $1.10 in 1982.

Conclusion

It would appear that the subsidies given to French poultry farmers are not the main cause of the differences in costs between the poultry farmers in the two countries. The importance of the regulations introduced in 1981 is now clear. Once the monetary compensatory amounts were reduced, the cost differences between poultry farmers in the United Kingdom and France became more apparent and a large increase in imports into the United Kingdom was likely. Import tariffs could not be imposed under the terms of the Treaty of Rome and the government in the United Kingdom was not prepared to assist poultry farmers to adopt new methods of production. Instead, the United Kingdom used health and sanitary regulations in an attempt to isolate its market from French competition.

This action highlighted an inconsistency in the Treaty of Rome and subsequent Community legislation. Production subsidies violate the spirit of the Treaty but they are not explicitly illegal as are tariffs. A subsidy, however, can alter the competitive balance as effectively as an import tariff. Community legislation provides no recourse against member states which indirectly subsidize production. Community members who choose to retaliate must either adopt subsidies themselves, or employ an alternative beggar-my-neighbor policy. The introduction of a countervailing duty, while subsidies are negotiated through bilateral consultation, would better serve the goals of the Community customs union in three ways.

First, a countervailing duty can be set at a level that exactly offsets the encroaching subsidy or similar trade distortion, thus reducing the level of resource misallocation.[5] Second, countries that do not subsidize production need not be penalized as they are currently when nontariff measures are used. A return to tariff protection would, therefore, minimize welfare losses from protectionist policies. Finally, the restoration of "justifiable" tariffs would at least increase a nation's accountability for its protective policies by quantifying protection at specific rates. Community member states, and other nations as well, have replaced concessions made in tariff negotiations with more obscure trade distortions. A return to tariff protection would restore national responsibility by replacing subdued measures of protection with those that are obvious. If nations at present show no enthusiasm for policies oriented towards free trade, limited tariff protection at least provides the most efficient and responsible alternative and in this way provides an intermediate step toward the realization of Community free trade objectives.

Notes and Sources

1. Material in this chapter is adapted from Anthony C. Crooks, *Case Study in Protectionism: British and French Poultry Trade, 1978–1982.* Unpublished Master's Thesis, Department of Agricultural Economics, University of Arizona, 1984.

2. *"Poultry Chairman Welcomes Turkey Ban,"* *Agra Europe*, 875, May 2, 1980, p. 6.

3. "UK Poultry Health Move Will Mean End to Imports from Most Countries," *Agra Europe*, 943, August 28, 1981, p. 2.

4. "EEC Court Rules Against UK in Newcastle Disease Case," *Agra Europe*, 983, July 15, 1982, p. 6.

5. Appendix E sets out the arguments for the use of a countervailing duty as opposed to a nontariff trade measure to demonstrate why the former is a more efficient policy instrument for offsetting production subsidies.

8

Future Prospects for Technical Trade Barriers in Agriculture

It has been shown in earlier chapters that nontariff measures have proliferated as successive multilateral trade negotiations within the GATT have all but eliminated protective tariffs. The domestic agricultural policies pursued by the industrialized countries and, in particular, by the United States, the European Community, and Japan, make it likely that nontariff measures will continue to be used as countries try to find ways of increasing their exports of surplus agricultural products and of reducing imports of other agricultural products.

Measurement of Protection

Nontariff protection is difficult to measure because protective devices are often concealed in administrative practices which have a large discretionary component. In earlier chapters it was demonstrated that protection of agriculture through the use of nontariff measures is recognizable in some instances and deliberate discrimination against foreign production can be identified. A rigorous evaluation of the effects of such measures is, however, extremely difficult.

A document published by UNCTAD in 1985[1] discussed various ways of measuring the scope and economic impact of nontariff barriers. The simplest of these was a frequency distribution or tabulation of the number of nontariff barriers in operation. To obtain an estimate of the quantitative impact, however, the authors turned to commodity analysis—measurement of the price and/or quantity changes in traded commodities. They recognized, however, that normal variations in trade flows and in macro-economic variables make precise evaluation impossible. In other words, only barriers which have a substantial impact on

123

on trade flows could be quantified through analyses of changes in price, supply and demand.

More specific to the measurement of nontariff barriers in agriculture is the concept of producer subsidy equivalents and consumer subsidy equivalents. This concept was developed by Professor Timothy Josling with the help of others in the early 1970s;[2] it is now widely used in technical discussions of trade agreement tactics. Producer subsidy equivalents are the cash equivalents of the various aids to a particular farm commodity—in other words, they are similar to an estimate of the effective tariff. Their merit lies in the fact that the use of producer subsidy equivalents makes possible comparisons of the level of support for different commodities in different countries. A weighted estimate of the aggregate level of support can also be calculated. The possibility of reducing the level of producer subsidy equivalents, through negotiation, in much the same way as tariffs are reduced, makes the concept attractive to trade negotiators.

A comprehensive review of past negotiations in the GATT and the problems and calculations of producer subsidy equivalents for most OECD countries plus selected less-developed countries was released in 1987 by the USDA.[3] A similar study by the OECD also used producer subsidy equivalents to calculate levels of protection.[4] Several other studies—by the World Bank and the International Food Policy Research Institute—provide analyses in this area but in these instances more emphasis is given to the less-developed countries. The main point is that a quantitative base now exists which would enable reductions in the level of agricultural support to be negotiated, if the will exists.

It should be pointed out, however, that producer subsidy equivalents are not a measure of the economic impact of the total subsidy or aid given to a particular agricultural commodity. Producer subsidy equivalents do not take account of possible changes in the domestic and international supply and demand schedules that have, or will result from changes in policy measures.

National and International Standards

A distinction should be made between nontariff barriers which are meant to restrict trade and others which have a different purpose. There is in existence a whole range of national statutes and regulations designed not only in the interest of consumers in particular countries but also in the interest of the international economy. The prohibition of foods unfit for consumption and of articles dangerous to the health of human beings, animals or plants, all fall into this category. Phytosanitary and other regulations perform a necessary service but discrepancies and redundancies among various national codes impose unnecessary costs on the expansion of trade.

The standards elaborated by the CODEX commission (see Chapter 4) have as their aim the health of all consumers and the maintenance of fair practices in the national and international food trade. They contain requirements which will ensure that the consumer can obtain a sound, wholesome, correctly labeled product free from adulteration. In order to ensure that all concerned have a chance to register their points of view, each standard goes through a system which involves ten separate steps. There is, however, no legal obligation on member governments to accept a recommended standard. If they do accept it, they do so in accordance with their own established legal and administrative procedures and in respect of the distribution of all products complying with the standard, whether the products are imported or home-produced.

On the other hand, while standards and regulations are not undesirable in themselves, those which discriminate against particular goods and services cannot be justified. Assertions that discriminatory intent and the effects of discrimination are difficult to judge, or that they are too difficult to measure, are not adequate reasons to delay the eradication of discriminatory nontariff barriers. Improved techniques of language translation, of communication and of assessment, moreover, assist in the analytical processes.

Nevertheless, while everyone would agree that certain regulations are necessary and desirable, legitimate differences of opinion arise over how restrictive these regulations should be. It is a tenet of economic science as practiced in most English speaking countries that the law should place as few impediments in the way of informed individual choice as possible, since the unfettered price mechanism is capable of accommodating the value people impute not only to their appetites, but to their comfort and safety as well. By the logic of economics, protection from harm and fraud has a calculable value and that value should be determined by careful weighing of its benefits against its costs. With this approach, societies would discover that they do not need the full measure of protection that some laws currently afford them, that they would choose to dispense with some of it to save resources that would otherwise be wasted; in other cases the public might deem more protection worthwhile. Cost-benefit analysis is a wise approach, but it does not resolve the question of risk tolerance levels completely. Scientists, engineers, courts, politicians and economists have all adopted their own conventions for dealing with uncertainty. Addressing regulatory reform through technical committees of diverse professional composition would bring all of them together, but then they would face the challenge of reconciling their respective standards of risk aversion in a coherent way that will appear to represent the public interest.

There has tended to be a proliferation of international regulatory bodies in the agricultural sector which implies in practice a dispersal of effort on the part of national states and other actors. It has been pointed out that many groups which are active at the national level in the consideration of regulatory measures lack the resources to take part in international deliberations.[5]   Also, the structuring of international institutions around the principle of member-state representation and of the soverign equality of states creates special problems for industry and other non-state participants.   When the international agencies lack enforcement power some governments are likely to be deterred from taking part either in the deliberations of such agencies or in adopting any suggested regulatory measures. The history of international organizations has been a source of frustration to those taking part.   One observer, however, sees some useful purposes in the frustration and failure in that:

(a) one set of institutional arrangements can lead to, or even create, others which may be an improvement;

(b) the accumulated weight of a resolution or recommendation as it moves in different forms through different organizations—as did those on the international trade in toxic chemicals during the late 1970s and early 1980s— compensates to some extent for any inherent lack of political visibility each may have.[6]

Because of the difficulties in measuring and quantifying their impact, nontariff measures do not lend themselves as easily as tariffs to bargaining in a multilateral context. The great interest in producer subsidy equivalents reflects a frantic groping for some such tool, albeit imperfect. It has been suggested that the success of the Kennedy Round multilateral negotiations was facilitated by each government's knowledge that it could, if necessary, substitute nontariff methods of protection for the explicit tariff protection it was giving up.[7]   No substitutes seem to be available for nontariff measures, which probably means that future negotiations must deal directly with government attitudes toward protection.

Harmonizing Technical Barriers

In the meantime, the problem of harmonizing the technical barriers to trade is receiving a good deal of attention. The problem arises in a most acute form in relation to sanitary and phytosanitary regulations.[8] Among the many objectives of the Uruguay Round of multilateral negotiations is that of "harmonizing the adverse effect that sanitary and phytosanitary regulations and barriers can have on trade in agriculture, taking into account the relevant international agreements." The preliminary negotiationg positions of all major participants called for harmonization and the role of existing international agreements and of standards organizations was recognized. What it also reflects, however, is the lack of an effective procedure within the GATT for the settlement of disputes in the area of phytosanitary and sanitary restrictions.

The United States has taken the lead in the Uruguay Round in endeavoring to harmonize health and sanitary regulations among countries. It has put forward a proposal the objectives of which are: (i) to harmonize health and sanitary regulations; (ii) to base domestic regulations on internationally agreed standards; and (iii) to base processes and production methods on equivalent guarantees.

In order to implement these objectives, an expansion of the rules and procedures of the GATT which govern technical barriers is proposed in order that these rules and procedures might: (i) apply more explicitly to processes and production methods; (ii) give greater recognition to the principle of equivalency of laws and regulations; and (iii) provide procedures for early technical and policy consultations on legal and regulatory changes that have a high potential for disrupting trade.

The United States has been careful to point out that in its proposal "harmonization" does not imply identical or uniform laws, but only acceptance of standards which first, provide substantially equivalent protection; second, vary within technically reasonable limits; and third, have been recognized by relevant international standards organizations. Such harmonization would include standards for all agricultural products, food, beverages, forest products, fish and fish products.

It has been pointed out that the American proposal appears to go well beyond defining a dispute settlement procedure, an alleged weakness of the Standards Code.[9] The proposal calls for "an internationally agreed approach" to the development and implementation of standards regulating trade in food and agricultural commodities. International standards would serve as the basis for "new technical requirements" and or the import and distribution of foods products. A "formal link" with international standards organizations is proposed. These organizations would serve two roles: (i) to judge the technical or scientific rationale of standards called into question by trading partners; and (ii) to determine the equivalency of standards.   The second role would be crucial to successful implementation of any agreement on the "principle of equivalency of standard laws and regulations."   The penalty that could be applied if a dispute could not be resolved is based on Article XXII and Article XXIII of the GATT.   More flexibility and a greater range in the response would be allowed under the proposal than under the Standards Code, which seems to allow suspension of concessions and obligations contained within that code.

At the heart of the American proposal is the principle of equivalency of laws and regulations on standards.   Within the United States it might happen that two production processes, or alternatively, product-testing procedures, would be needed for the domestic and international markets.   The American proposal requires the international recognition that food standards based on processes and production methods and those based on micro-biological standards can provide equivalent protection in terms of health and food safety.   This equivalency has been the source of domestic disputes between processors and consumers in the United States.

If it is assumed that agreement could be reached on the principle of equivalency, implementation would require either a procedure for determining equivalency within the rules and procedures of the GATT, or a procedure to settle disputes arising over equivalency.   The appeal of the former would be its uniform application; the "disadvantage" is the huge amount of rule-making that could be forced upon the GATT Secretariat.   The appeal of the latter is that only when equivalency is called into question will the GATT become involved in rule making.   Despite the disparity of interests, progress on the harmonization of sanitary and phytosanitary regulations is more promising than in other areas of the negotiations currently taking place in the GATT.

Agriculture in the Uruguay Round

A proposal was put forward by the United States in the early stages of the Uruguay Round of regulations for all trade-distorting agricultural subsidies to be eliminated over a ten-year period.   The original proposal which was tabled in 1987 was revised in 1989 and is the subject of current (1990) negotiations along with revised versions of the proposals from the European Community, the Cairns Group, Japan, and other countries.   Because the American proposal and the American position have led to so much controversy, a complete text of the most recent (December 1989) American position under the title *Submission of the*

United States on Comprehensive Long-term Agricultural Reform is presented in Appendix A. The proposal put forward by the European Community emphasized market stabilization as part of a process to reduce its onerous agricultural budget to a more manageable level. The Cairns Group of developing and developed nations sought a freeze on current trade distorting measures, plus an agreement to reduce levels of support over a ten-year period. A Canadian proposal went one step further by calling for eventual elimination of all subsidies, within an interim target of five years. The Nordic countries wanted immediate action to prevent increases in surplus products, an initial freeze followed by reductions in export subsidies and reduced barriers to market access. Japan proposed a freeze on export subsidies which would subsequently be phased out, a reduction in the domestic subsidies which lead to trade distortions and negotiated improvements in market access.[10]

While some of the elements in these proposals are similar, others are not compatible; in some instances, the United States and the European Community started the Uruguay Round at opposite ends of the spectrum. One observer, Dale Hathaway, has pointed out[11] that it is not possible to have free domestic markets and government-managed international markets; nor is it possible to have free international markets and tightly controlled domestic markets for agricultural products. Domestic and international markets are closely interlinked. The same observer also pays much attention to the political realities of the agricultural sectors of the major trading groups. He urges that negotiations should begin with agreement on "general issues or rules to be applied to various types of import controls and subsidies, leaving the issues of applying principles . . . until the end."[12]

As of mid-1990, many revised and some new proposals had been tabled at the GATT Negotiating Group on Agriculture. They include, among others, those of the United States, the European Community, the Cairns Group and Japan.[13] Those of the United States, the European Community, the Cairns Group and Japan will have the greatest weight in the negotiating process for the balance of the Uruguay Round-1990 and beyond.

It will be possible here to state only the main objectives of the proposals of these major groups and to make a few brief comparisons between them.[14] Since the outset of the Uruguay Round the principal concern of the United States has been to achieve meaningful long-term agricultural policy reforms which would lead to market orientation of world agriculture. Hence, the central theme of its latest proposal is to ". . . guide agricultural production and trade toward a market oriented system governed by strengthened and more operationally effective

GATT rules and disciplines, and to integrate agriculture fully into the GATT. "

The European Community has always been more concerned with market stability than has the United States. Thus, stability along with food security and distributional equity are its principal concerns. In its latest proposal, these would be achieved through various techniques which would "progressively reduce support to the extent necessary to re-establish balanced markets and a more market oriented trading system—not to set *a priori* and *in abstracto* a final level of support." The Cairns Group wants immediate short-term measures which would lead to long-term fundamental reform. The ultimate goal is—"a competitive, efficient and market responsive world agricultural system [which] would serve the common interest of developed and developing countries alike." Japan's major concern has always been and still is food security. The main objective of its proposal is therefore to ensure that "—full consideration should be given to the special nature of agriculture which is constrained by land and climatic conditions, and to the multifarious roles played by agriculture such as food security." Japan wants this assurance within any process which would make "substantial reductions in agricultural support and protection."

The Mid-Term Agreement for agriculture approved by the GATT Trade Negotiations Committee on April 7, 1989 (MTN,TNC/11), states that, no later than the end of 1990, participants will agree on a long-term agricultural reform program and a period of time for its implementation. Also, the long-term objective is to provide for substantial progressive reductions in agricultural support and protection, all of which would be sustained by strengthened and more operationally effective GATT rules and disciplines. The United States is insisting that trade reform be achieved by attacking the problems through a subset of policy instruments categorized as: market access, export competition, internal support and sanitary and phytosanitary measures, (see Table 8.1).

Rigorous negotiating sessions began on these latest proposals from GATT members. Even though December 1990 is the target date for an agreement on long-term agricultural policy reform, there is no guarantee that negotiations will not be extended, especially since the United States will be wrestling with its farm bill legislation, and the European Community will be dealing with the economic fallout from recent events in Eastern Europe. Negotiating positions of certain participants have moved significantly since 1987, but in some cases requests have become more demanding. Optimists[15] perceive that there has been a remarkable degree of flexibility and that negotiating positions have narrowed. Such optimism, whether politically justified in light of history,

TABLE 8.1 Submission by the United States on Comprehensive
Long-term Agricultural Reform

Market Access
    Objective: Substantial progressive reduction in all import
                 protection

Export Competition
    Objective: Elimination of export subsidies and export prohibitions

Internal Support
    Objective: Development of new GATT rules and disciplines
                 covering all trade–distorting subsidies leading to the
                 elimination of the most trade–distorting policies

Sanitary and Phytosanitary Regulations and Barriers
    Objective: Establishment of an international process for settling
                 trade disputes involving food safety, animal health and
                 plant health issues and for promoting harmonization

---

is built on the thesis that nations will now act on those agricultural and trade reform measures which in the past have been delayed or rejected. The author of this book is among the doubters. Witness the outcome of the Houston 1990 Economic Summit.

Most participants and observers expect some compromise to emerge from the Uruguay Round of negotiations in part because of the political timetable of the major parties. The United States is considering a farm bill, and events in Eastern Europe have altered political and economic affairs. While several of the principals are driven by financial consideration to seek a reduction in the budget costs of agricultural support, that pressure was not as great in 1990 as it was earlier in the negotiations.

Agricultural subsidies in the industrial market economies amount to enormous sums. As recently as 1986 consumer costs were $47 billion in the European Community, $38 billion in the United States and $34 billion in Japan (Figure 8.1). Producers receive substantially less of sums but the costs are nevertheless very high. Budgetary costs to the industrial economies dropped considerably after 1988 due to rising prices of agricultural commodities. In Japan, even though budgetary costs for agriculture, forestry, and fishing are declining, the combination of agricultural and tax policies permeate the entire economy. High

132

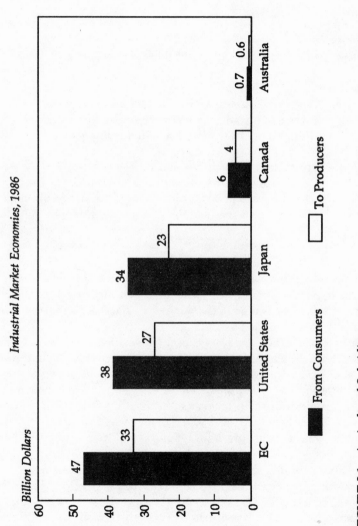

Industrial Market Economies, 1986

FIGURE 8.1. Agricultural Subsidies.

Source: Derived from: Vernon O. Roningen and Parveen M. Dixit, *How Level is the Playing Field? — An Economic Analysis of Agricultural Policy Reforms in Industrial Market Economies,* Foreign Agricultural Economics Report No. 239, Agriculture and Trade Analysis Division, ERS, USDA, December, 1989.

internal prices for rice with much farm land devoted to rice production results in high prices for land and high costs for other Japanese farm products. High land prices contribute to a housing shortage (or high prices for housing) while high food prices absorb much Japanese purchasing power. These two conditions reduce the Japanese demand for other consumer goods, including imports. It has been estimated[16] that Japanese consumers and taxpayers paid about as much for their agricultural support as they would have paid for food if world prices had prevailed within Japan, a total of about $13 billion.

Role of the GATT

The question has been raised whether negotiations on changes in national agricultural policies should take place in the GATT. There are those who hold that the GATT is an institution where negotiations on legal, regulatory and modality matters should take place; it is not a forum for designing and assessing the effects of national agricultural policies. On the other hand, those supporting a more active stance on policy design and assessment admit that while the legitimacy of domestic policy objectives is not questioned, the measures used to achieve such objectives must be brought under the supervision of the GATT if they create international market distortions.

Critics of the GATT could point out, however, that the supervisory powers are of little practical value if the agency has no power to insist on changes in domestic agricultural policies which distort international trade in agriculture.

The fact that the GATT does not have the power to enforce its rules and procedures has been a weakness since its inception.[17] Currently, however, the accumulation of grievances without solution, particularly against the large trading blocs, has increased the tenor of the protests. The diversity of nontariff barriers now being used as part of domestic agricultural policy makes it evident that a continued surveillance by a body with enforcement power must be created or the powers of the GATT must be greatly strengthened. The need is for an agency which will have both expert advice and enforcement powers if changes are to be brought about.

Whatever the results of the Uruguay Round deliberations, it is likely that there will be an increased demand for technical experts. They will be needed first, to give advice during negotiations on technical matters such as alternative health, sanitary, processing, packaging or other regulations. Secondly, they would be needed to provide impartial technical information on the validity of proposals by groups of producers, exporters or importers, or by the government regulators themselves. When trade disputes arise, or threaten to arise, an international panel of experts could provide scientific information on production, processing, marketing or use that can prevent specious arguments and perhaps encourage the political leadership to make rational choices. Any examination of recent trade disputes or of the list of existing non-tariff barriers will illustrate the variety of technical information needed to understand fully some current trade issues. While a panel of experts in the several fields may not always agree, they are very likely to narrow the range of differences and to help political and administrative authorities to focus on the central issue, as well as providing a possible means of bringing the issues to public attention.

Experts are also needed by GATT-participating countries in a policy-monitoring context. One group which is involved in the current scene is the *International Policy Council on Agriculture and Trade* which issues periodic communiqués on the GATT and similar subjects. It is instructive to cite here points made in a communiqué on GATT and the Uruguay Round of April 28, 1990 but which also contains key elements for continued sound trade and agricultural policies.[18]

a. A general commitment to a significant reduction in the level of distorting support and protection. Some aggregate measures of support should be used for setting this target and monitoring progress.   The reductions should in principle apply to all countries and commodities equally, though some flexibility may have to be accepted.

b. A reduction in import barriers by an amount at least comparable to that agreed for the overall level of trade-distorting support and an introduction of progressive tariffi-cation.   All waivers and similar exemptions would be removed as a part of the reinstrumentation of import barriers.

c. The removal or reduction of export subsidies by an amount at least comparable to that agreed for the overall level of support and, so long as they exist, their subjection to special disciplines.

d. The classification of domestic subsidies depending on their trade-distorting nature. Those which satisfy certain agreed criteria would be permitted.   Others would either be prohibited or be subject to a phased reduction at the same rate as the overall commitment.

e. The need, in the interest of achieving an agreement, to deal with a number of specific trade and non-trade issues as special cases. These might include the rebalancing of EC support policy and the appropriate recognition of concerns about food security with a prohibition on export restrictions in times of scarcity.

f. A recognition of the special needs of developing countries and the former centrally planned economies so that they can participate fully in the GATT process. Developing country exporters have been adversely affected by the present disarray in world agricultural and tropical product trade. They can therefore expect greater discipline to be applied to developed country trade policies, thus promoting growth and alleviating the problems of the heavily indebted countries.

    If trade liberalization raises world food prices or reduces food aid levels, thus harming low income food importers, the means of easing these problems should be negotiated in the present Round. This would be in addition to development aid to facilitate more rapid economic growth; agricultural development strategies to enhance on-farm and off-farm employment opportunities; and adequate incentives to farmers to increase badly needed food production.

g. A revision of GATT rules to underpin the commitments entered into, to achieve transparency and to ensure effective discipline on all Contracting Parties. In addition, changes are needed to effectuate commitments already mentioned, as well as to develop regulations in the health and phytosanitary area (Article XX) and to strengthen processes for dispute settlement. GATT resources and procedures should be strengthened to ensure adequate monitoring, implementation and enforcement of the agriculture agreement.

The Council emphasizes that a successful outcome for agriculture in the Uruguay Round is essential for a more rational world agriculture. Developed and developing countries alike have large interests at stake. All parties in earlier phases of this Round agreed to substantial and progressive reduction in each area of the negotiations—internal support, export subsidies and barriers to market access. Toward the end it was unilateral and bilateral actions and the pursuit of narrow sectoral interests that continued to impede progress toward overall agreement. The expert advice of the Council, like that of all experts, is greatly needed if protectionism is to be reduced.

Finally, the removal of nontariff barriers is likely to have little meaning to the general public or even to many financial reporters, except in a few highly publicized cases (such as a "chicken war" or a substantial meat quota increase by Japan). The removal, or halving, of a tariff rate is much more easily understood and therefore a more glamorous accomplishment to a trade negotiation. The many rounds of trade negotiations which have taken place in the postwar period have eliminated most of the important issues in relation to tariff levels; what remain are a series of much less tractable measures: quotas, health and sanitary regulations, rules dealing with production processes. Safe food has become a leading issue in this group. A recent study states that food safety regulations may pose some of the most serious obstacles to agricultural trade in the decade ahead.[19] Such obstacles will continue to be rather intractable until western trading countries can resolve the conflicts that continue to stimulate greater production even while many farmers remain poor.

Summary and Conclusions

In this chapter, some of the obstacles that stand in the way of freer trade in agricultural products and the developments which are taking place have been briefly considered. The following paragraphs summarize the main points:

(a) Nontariff barriers have grown in number and variety over the period during which tariffs have been reduced. It has been argued that this is a cause and effect relationship. Efforts to limit the growth and the application of nontariff barriers both through the GATT and through bilateral discussions have often proved ineffectual or have been hindered through the use of delaying tactics.

(b) Precise measurement of the protectionist element of nontariff measures is difficult. In many instances, only crude measurements are possible; only those technical barriers which have a major impact on trade are capable of being measured with some precision. The concept of the producer subsidy equivalent, however, is proving to be a useful negotiating tool, at least in the traditional framework of trade negotiations.

(c) It has become very obvious that no real progress in reducing nontariff barriers will be possible unless substantial adjustments are made in the domestic agricultural policies of the three major participants: the United States, the European Community and Japan. The current Uruguay Round negotiations may have no more success in this area than the earlier multilateral deliberations.

(d) Taxpayers and consumers have a very substantial interest in a successful outcome of the current Uruguay Round negotiations. Reduced food costs for the consumer will allow more expenditure on other goods and services; reduced budget outlays for governments will facilitate lower taxes or high expenditure on, for example, social projects. The problem will be how to mobilize consumer opinion and how to compensate for the adverse effects on the farming community.

(e) Negotiations for the reduction or elimination of technical barriers to trade in agricultural products means that negotiators will require more scientific and technical information. It will also be important to have at least some estimate of which technical barriers lead to the greatest distortions in trade.

(f) The GATT must have power to enforce decisions upon even its most powerful members although some appeal process should also be available. The question arises, however, as to whether the GATT is the appropriate organization to take on the new responsibilities that many hope will emerge from the Uruguay Round.

Notes and Sources

1. Alan Deardorff and Robert M. Stern, *Methods of Measurement of Nontariff Barriers*, UNCTAD/ST/MD/28, (Geneva: UNCTAD Secretariat, 1985).

2. Timothy Josling and Jimmye Hillman, *Agricultural Protection and Stabilization Policies: A Framework of Measurement in the Context of Agricultural Adjustment*, C75/LIM/2. (Rome: Food and Agriculture Organization, 1975). This document built on the earlier FAO study (C73/LIM/9) by Josling *et al.*, *Agricultural Protection: Domestic Policy and International Trade*, (Rome: Food and Agriculture Organization, 1973) which was an attempt to develop a methodology for estimating domestic and trade effects of government intervention in agricultural markets illustrated with reference to selected countries.

3. *Government Intervention in Agriculture: Measurement, Evaluation and Implications for Trade Negotiations*, Economic Research Service, USDA, 1987.   Also *Estimates of Producer and Consumer Subsidy Equivalents: Government Intervention in Agriculture*, 1982-86, Economic Research Service, USDA, 1988.

4. *National Policies and Agricultural Trade* (Paris: Organization for Economic Cooperation and Development, 1987).

5. For an excellent discussion of this issue with respect to pesticides see Robert Boardman, *Pesticides in World Agriculture: The Politics of International Regulation.* (London: Macmillan, 1986.) I am grateful to Dr. Boardman for his permission to cite his work in this paragraph.

6. *Ibid*, p. 180.

7. Brian Hindley, "Negotiations on Nontariff Barriers," Chapter 6 in *Towards an Open World Economy* (London: Macmillan, 1972), p. 127.

8. Maury E. Bredahl and Kenneth W. Forsythe, "Harmonizing Phytosanitary and Sanitary Regulations," *World Economy*, Vol. 12, No. 2, June 1989, pp. 189-206.

9. *Ibid.*

10. M. Ann Tutwiler and George E. Rossmiller, *Prescriptions for Success in the GATT*, (Washington: National Center for Food and Agricultural Policy, Resources for the Future, 1987) and *Summary Report, Negotiating a Framework for Action*, prepared under the guidance of the chairman of the International Agricultural Trade Research Consortium, Professor David Blandford, Cornell University, Ithaca, New York, 1988.   Two other summaries were entitled: *Assessing the Benefits of Trade Liberalization* and *Designing Acceptable Agricultural Policies*, unnumbered, undated, position papers.   For copies contact Professor David Blandford, Cornell University, Ithaca.

11. Dale Hathaway, *Agriculture and the GATT: Rewriting the Rules*, Institute for International Economics, Washington, D.C., Sept. 1987, p. 133.

12. *Ibid.*, p. 130.

13. The proposal of the United States has already been cited as *Appendix A*.   Following are the titles and official citations of the European Community, the Cairns Group and Japan.

GATT.   Multilateral Trade Negotiations: The Uruguay Round (MTN), Group of Negotiations of Goods (GNG):

"Negotiating Group on Agriculture.   Statement by Japan," MTN.GNG/NGS/W/104, October 2, 1989.

"Comprehensive Proposal for the Long-Term Reform of Agricultural Trade," (Cairns Group), MTM.GNG/NG5/W/128, November 27, 1989.

"Global Proposal of the European Community on the Long-Term Objectives for the Multilateral Negotiation on Agricultural Questions," MTN.GNG/NG5/W/145, December 20, 1989.

GATT.   "Submission of the United States on Comprehensive Long-Term Agricultural Reform," [1989].

14. I am grateful to Dr. Kelly White of the Economic Research Service, USDA, for his assistance and that of his staff in making these summary comparisons of documents–proposals which comprise great length and detail.

15. Stefan Tangermann. "Options and Prospects: A Feasible Package," in *Agriculture in the Uruguay Round of GATT Negotiations: The Final Stages*, Department of Agricultural Economics and Business, AEB 90/2, University of Guelph, May 1990.

16. D. Gale Johnson, Kenzo Hemmi and Pierre Lardinois, *Agricultural Policy and Trade: Adjusting Domestic Programs in an International Framework*, Triangle Papers No. 29. (New York: The New York University Press, 1985.)

17. See, for example, Robert E. Hudec, *Adjudication of International Trade Disputes*, Thames Essay No. 16, (London: Trade Policy Research Centre, 1978).

18. International Policy Council on Agriculture and Trade, a communiqué entitled: "GATT Negotiations Told: Agreement Is Clearly Within Reach," 1616 P Street, N.W., Washington, DC, April 28, 1990.

19. Raney, Terri, and David Kelch. "The Safe Food Issue: New Nontariff Barriers?" *World Agriculture Situation and Outlook Report*, WAS–59. Washington, D.C.: Economic Research Service, USDA, June 1990, pp. 28-32.

Appendix A

Submission of the United States on Comprehensive Long-Term Agricultural Reform

The Mid-Term Agreement for Agriculture approved by the Trade Negotiations Committee on April 7, 1989 (MTN.TNC/11) states that:

- not later than the end of 1990, participants will agree on a long-term agricultural reform program and the period of time for implementation;
- the long-term objective is to provide for substantial progressive reductions in agricultural support and protection, sustained over an agreed period of time, resulting in correcting and preventing restrictions and distortions in world agricultural markets;
- the strengthened and more operationally effective GATT rules and disciplines, which would be equally applicable to all Contracting Parties, and the commitments to be negotiated, should encompass all measures affecting directly or indirectly import access and export competition;
- proposals to achieve these objectives are to be submitted by December 1989.

In accordance with these guidelines, the United States hereby submits to the Contracting Parties a comprehensive proposal for agricultural reform. The comprehensive nature of this proposal recognizes the wide variety of internal and border measures employed across countries and commodities, and the complex interaction among these policies. The broad array of policy instruments is categorized into four subsets: import access, export competition, internal support and sanitary and phytosanitary measures. The United States proposal calls for reform in all four areas, and these reforms should be viewed as integral parts of a comprehensive package and not as four separate proposals. Our comprehensive proposal is designed to guide agricultural production and trade toward a market-oriented system governed by strengthened and more operationally effective GATT rules and disciplines and to integrate agriculture fully into the GATT.

The United States proposal is designed to correct and prevent the many problems and distortions of current agricultural policies. These include costs to consumers and taxpayers that exceed $275 billion annually; incentives for overproduction; subsidized disposal of surpluses; internal support systems that distort trade and inhibit market development; and import barriers that misallocate resources, reduce the level of food purchases, and limit consumer choice.

Less evident failures of current policies also are becoming an increasing concern. Costs involved in disposing of surplus production have instilled in governments and producers a fear of promising new technologies. High price supports that encourage production at the expense of sound management have led to excessive use of chemical fertilizers and pesticides, heightening social concerns over water quality, food safety, and other environmental issues. Moreover, poor land management, soil erosion, and deforestation are too often the results of today's highly subsidized agricultural systems.

Under the United States proposal, the structural rigidities of present systems would be gradually replaced by a more market-oriented environment in which farm income and other objectives could be achieved at far less cost, program benefits would be distributed more equitably among farmers everywhere, and opportunities for market growth would be enhanced. Producers would be free to choose the mix of production that best fits their resource base, unshackled from policies that discourage efficiency and new product development. Marginal land and land damaged by soil erosion, deforestation or poor land management could be improved by reforestation or conservation programs. Developing countries would benefit from increased trading opportunities. *Bona fide* food aid would not be affected under the proposal. The continuation of food aid programs would help ensure an adequate supply of food for developing country importers. Food security would be enhanced through the elimination of export restrictions and prohibitions for reasons of short supply and the reform of other trade-distorting policies. Such reforms would increase production efficiency and the availability of food for all citizens of the world.

To achieve a fair and market-oriented agricultural trading system, the United States proposal encourages the transformation of many existing internal support measures to policies that will be less trade-

distorting. Traditional forms of support that are directly tied to production and price levels would be phased out over a ten-year period.
Certain other policies that are less often abused but still capable of significant distortions would be reduced over the ten-year period through
the use of an aggregate measure of support and subjected to GATT
disciplines. Minimally trade-distorting policies would be permitted.

All nontariff import barriers would be converted to tariffs and,
along with preexisting tariffs, would be reduced over time. Export subsidies would be phased out over five years. Export restrictions imposed
for short-supply reasons and export tax differentials would be eliminated. For sanitary and phytosanitary measures, disciplines are proposed
that would establish new procedures for notification, consultation, and
dispute settlement. These procedures would require the use of sound
scientific evidence in the development of health-related measures and
would recognize the principle of equivalency in order to prevent the imposition of unwarranted trade barriers. Respected international scientific organizations would be asked to provide standards and scientific
evidence to enhance settlement of disputes.

Agricultural trade reform requires the active participation of all
Contracting Parties. Strengthened and more operationally effective
GATT rules and disciplines should be applicable to all Contracting
Parties. At the same time, the United States understands the distinctive needs of developing countries, particularly with respect to infrastructure. Developing countries may need longer timeframes for adjustment. As indicated in the Punta del Este Declaration and confirmed in
the Mid-Term Agreement, contributions of developing country to the
negotiations should reflect their individual levels of economic and
agricultural development.

PRODUCT COVERAGE

The United States proposes that the rules, disciplines and transition
mechanisms outlined in this paper apply to the commodities listed in
Annex 1.

IMPORT ACCESS

Objective
The objective is to orient domestic production to market forces
through conversion of all nontariff barriers to bound tariffs and
ultimately reduce all agricultural tariffs to zero or low levels. After an
agreed transition period, all import protection would be in the form of
tariffs.

Rules and Disciplines
The United States proposes that:

- all waivers and derogations, protocols of accession and grandfather clauses that allow derogations to existing GATT rules for import access be eliminated;
- variable import levies, voluntary restraint agreements, minimum import prices, and other import barriers not explicitly provided for in the GATT be prohibited;
- GATT Article XI: 2(c) be eliminated.

In addition, Contracting Parties would discontinue the use of restrictive import licensing and other import barriers prohibited under current GATT rules. All disciplines would be applied equally to marketing boards and other state-trading organizations. Article XVII should be strengthened along the lines proposed by the United States in the Negotiating Group on GATT Articles.

Implementation
A. *Tariffs.* All tariffs, including those resulting from the conversion of nontariff barriers to tariffs, would be bound on January 1, 1991, and then reduced over a ten-year transition period to final bound rates to be negotiated. The final tariffs would be at zero or low levels.
B. *Nontariff Measures.* Nontariff barriers include policies such as quotas, variable levies, import restrictions or prohibitions administered in connection with marketing boards and state trading operations, voluntary restraint agreements, restrictive licensing practices, and other import restrictions and prohibitions.
As a first step in the process of liberalizing import access, no new nontariff measures would be permitted and existing nontariff barriers would be replaced with a tariff-rate quota on January 1, 1991.
Tariff-rate quotas will permit an orderly transition from the extremely high levels of import protection provided by some current nontariff barriers to a tariff-based import regime. They will ease the transition for importing countries, while ensuring some degree of minimum access for exporting countries. They will also permit an orderly phaseout of country-specific import quotas.
The initial quota for each commodity (tariff-line item) would be set at a level equivalent to (1) the level of imports existing in 1990 or some recent historical period, or (2) a negotiated level of imports in the case of import prohibitions or virtual prohibitions. Tariffs within the quotas would be bound at agreed upon rates.

Bound tariffs would be the only form of import protection on imports outside the quota. In our July discussion paper of tariffication, we proposed that the initial tariff be based on the gap between world and domestic prices for each commodity (tariff-line item) where trade is now affected by nontariff barriers. Tariff rates could be expressed on an *ad valorem* or per unit basis and would be calculated on the basis of average prices for 1986–1988.

Liberalization of import access would be achieved by:

(a) a progressive annual reduction of over-quota tariffs to final bound rates; and

(b) expansion of initial quotas by agreed minimum amounts during the transition period.

At the end of the ten-year transition period, Contracting Parties would remove any remaining quotas and the final bound tariffs would be the only form of import protection.

Safeguard Mechanism

During the transition period, a special safeguard mechanism would operate to protect against import surges. The mechanism would have two trigger levels:

(a) If the previous year's imports of a particular commodity were less than three percent of domestic consumption, the safeguard would be triggered if imports in the current year exceeded 160 percent of the previous year's imports.

(b) If the previous year's imports of a particular commodity were equal to or greater than three percent of domestic consumption, the safeguard would be triggered if imports in the current year exceeded 120 percent of the previous year's imports.

A different safeguard mechanism may be needed for perishable commodities. Once the safeguard mechanism is triggered, a country would be allowed to revert back to a specified level of tariff protection for the remainder of the year. A shorter time period may be appropriate for perishable commodities. At the end of the year, the tariff snapback would be terminated and further tariff reductions would be implemented in accordance with the agreed schedule.

At the end of the transition period, safeguard actions will be permitted in accordance with Article XIX as revised during the Uruguay Round negotiations.

EXPORT COMPETITION

Objective

The objective is to orient more effectively domestic production to market forces through the elimination of all export subsidies and export prohibitions and restrictions on products covered by the Negotiating Group on Agriculture.

I. EXPORT SUBSIDIES

Rules and Disciplines

Contracting Parties would agree not to grant any form of subsidy on exports of the products listed in Annex 1. A proposed illustrative list of prohibited export subsidies is contained in Annex 2. At this point, the proposed list is the Illustrative List of Export Subsidies contained in the Subsidies Code. However, the United States reserves the right to propose amendments to this list to ensure that it is consistent with any changes that may be agreed upon in the Subsidies Negotiating Group and/or to ensure that it precludes export subsidies which could be specific to the agricultural trading system. Conforming amendments to the GATT instruments would need to be made to implement these new rules on export subsidies.

Only *bona fide* food aid would be exempt from this prohibition. However, the United States recognizes that there is a need to develop improved disciplines on food aid to ensure that such activities meet the needs of developing countries but do not distort normal commercial sales. New rules may need to be developed to govern the granting of food aid. The Contracting Parties would have to agree on guidelines which would clarify such issues as the conditions under which food aid may be provided, the categories of countries eligible, the kinds of commodities which could be provided and permissible terms, *i.e.*, what concessional arrangements would be acceptable.

Implementation

A five-year period is proposed for the phase-out of export subsidies. The basis for the phase-out could be either government expenditures and revenue losses in the base period, or quantities of commodities receiving export subsidy benefits.

II. EXPORT RESTRICTIONS AND PROHIBITIONS

Rules and Disciplines

Remove from GATT Article XI 2.(a) permission for GATT Contracting Parties to restrict or prohibit exports of agricultural food products to relieve short supply. This change was proposed earlier in a United States submission which addressed food security (MTN.GNG.NG5/W/61).

If a Contracting Party maintains export taxes, duties or charges on products that are used as inputs for the production of other products, and if such taxes, duties or charges are higher than the rate charged on the secondary products, then the differential between such taxes, duties or surcharges must be progressively reduced or eliminated. The purpose of this provision is to prevent countries from using a differential export tax structure to discourage exports of raw materials and thereby ensure a ready supply of artificially low-priced inputs for domestic processing industries.

Implementation

The change proposed in the first paragraph of the section above should be implemented in one step on January 1, 1991.

Elimination of the differential in export charges would take place on the same schedule as the phase-out of export subsidies (five years).

INTERNAL SUPPORT

Objective

The objective is to orient more effectively domestic agricultural policies to market forces through substantial progressive reductions in the trade-distortive elements of internal support policies.

General Approach

The Uruguay Round negotiations should address all domestic government programs, including programs of sub-national units of governments and government-sponsored organizations. The agricultural negotiations should focus on the wide range of domestic government programs that are unique to the agricultural sector.

To facilitate discussion of internal support programs, reference is made to general policy categories rather than specific policies. These general categories are intended to cover all internal support measures in the agricultural sector, including programs for producers, processors and consumers. The categories are listed in Annex 3. The categories are based on general characteristics shared by specific internal support measures. At this stage in the negotiations, general policy categories provide a sufficient basis to move forward. More precise attention to specific policies will be required at later stages in the negotiations.

There are wide variations among these policy categories with respect to the level of government support and the magnitude of trade distortion that exists. Although in the long-run nearly all policies that transfer resources to the agricultural sector have some effect on producer and processor incentives, and hence on output, certain methods of granting support have a significantly greater effect on the current agricultural trading system. These forms of support are directly tied to production or prices and should be phased out over the transition period. Support from other policies, less abused but capable of significant distortions, would be reduced and subject to GATT disciplines. On the other hand, policies that are minimally trade-distorting would be exempt from commitments to reduce support. This would allow Contracting Parties to use such policies to pursue national objectives such as resource conservation and environmental enhancement and development, while minimizing distortions in world agricultural trade.

Our three-tiered approach to domestic subsidy discipline can be summarized as follows:

1. Administered price policies;
2. Income support policies linked to production or marketing;
3. Any input subsidy that is not provided to producers and processors of agricultural commodities on an equal basis;
4. Certain marketing programs (*e.g.*, transportation subsidies);
5. Any investment subsidy that is not provided to producers and processors of agricultural commodities on an equal basis.

Policies to be disciplined:

> Other programs not elsewhere specified including, but not limited to, input or investment subsidies provided to any producer or processor of agricultural commodities on an equal basis and certain policies from categories listed elsewhere that do not meet the agreed upon criteria for permitted policies or policies to be phased out.

Permitted policies:

1. Income support policies not linked to production or marketing;
2. Environmental and conservation programs;
3. *Bona fide* disaster assistance;
4. *Bona fide* domestic food aid;

5. Certain marketing programs (*e.g.*, market information, most market promotion programs, inspection and grading);
6. General services (*e.g.*, research, extension and eduation);
7. Resource retirement programs;
8. Certain programs to stockpile food reserves.

Certain government programs which affect agriculture along with other sectors of the economy are more appropriately addressed in other Negotiating Groups. Examples include rural development programs that provide subsidies to all rural residents (small-town residents as well as farmers or processors of agricultural products) or government-funded highway systems.

Rules and Disciplines
New GATT rules must be devised that establish detailed criteria identifying policies to be phased out and permitted policies. Support from all policies that do not fit these criteria would be subject to specific disciplines as outlined below. Conforming amendments to the GATT instruments would need to be agreed upon to implement these new rules. The rules would be equally applicable to national and subnational policies.
A. *Policies to be phased out.* Any government program or policy meeting the following descriptions would be prohibited after a ten-year period:

- policies, other than border measures, that have resulted in or are designed to result in domestic prices higher than prices prevailing on the world market;
- income support payments to producers that do not meet the criteria for permitted policies (see below);
- subsidies on inputs or investments that are not provided to producers or processors of agricultural commodities on an equal basis;
- marketing subsidies that do not meet the criteria for permitted policies (see below).

New GATT disciplines should lead to the eventual elimination of these policies during a ten-year transition period. In order to ensure that the policies subject to these new prohibitions are carefully and unambiguously defined, the new GATT disciplines should be accompanied by a detailed interpretive note that would provide unambiguous definitions of the affected measures. In addition, provision should be made in the new rules for the periodic review and updating of the prohibited

B. *Permitted policies.* Government programs or policies meeting the following descriptions would be included in this category:

- direct income payments to producers that are not tied to current production, prices, the cost of production or marketing of agricultural commodities;
- programs for the development and implementation of *bona fide* conservation and environment protection plans and practices;
- disaster assistance keyed to *bona fide* production losses;
- *bona fide* domestic food aid;
- marketing programs that do not confer an economic benefit—in the form of price discounts, cash or in-kind payments, etc.—on the purchaser at any level of the marketing chain;
- general services that do not provide direct price or income support or subsidized inputs to producers, processors or consumers;
- programs to remove land or other production factors from agriculture or to facilitate the transition process; and
- programs for stockpiling food reserves that do not provide direct price or income support or subsidized inputs to producers, processors or consumers.

These policies would not be subject to the commitment to reduce support and protection agreed to at the Mid-Term Review. As with the list of policies to be phased-out, explicit criteria identifying permitted policies should be carefully defined in an interpretive note.

C. *Policies to be disciplined.* All policies that do not meet the criteria for permitted policies nor fit within the criteria for policies to be phased-out would be subject to specific GATT disciplines designed to prevent their use in ways that would nullify or impair concessions or cause serious prejudice or material injury to a Contracting Party.

In addition, Contracting Parties would negotiate reductions in support granted through this category of policies. Commitments in this regard would be expressed in terms of an agreed upon aggregate measure of support (AMS). These reduction commitments, together with the new GATT disciplines, should further reduce trade distortions and ensure that Contracting Parties do not simply transfer resources from one trade-distorting program to another as policies are phased out.

Implementation

Commitments on internal support would be implemented over a ten-year period.

A. *Policies to be phased out.* Contracting Parties would be free to choose the transition mechanism best suited to their particular policies. For example, Contracting Parties using administered price policies could progressively reduce either administered prices, or the amount of production eligible for price supports, or both. However, each Contracting Party will be required to choose a mechanism that will lead to reductions in equal annual steps over the transition period and culminate in the elimination of the policy in question. Commitments should be made by policy and, in most cases, by commodity as well.

B. *Policies to be disciplined.* An AMS approach would provide a convenient method of reducing support for these policies. The Producer Subsidy Equivalent developed by the OECD provides one such approach; others could be developed. The AMS level would be bound at progressively reduced rates over the transition period. The AMS calculation would include all types of government support that are not explicitly prohibited or permitted under new GATT rules. Since border measures would not be included, many of the methodological problems presently associated with AMS calculations could be avoided. Commitments would be implemented through a negotiated level of linear cuts in the AMS over the transition period.

SANITARY AND PHYTOSANITARY MEASURES

Objective

To provide a mechanism for notification, consultation and dispute settlement which would ensure that measures taken to protect animal, plant and human health are based on sound scientific evidence and recognize the principle of equivalency.

Rules and Disciplines

The United States proposes that Article XX(b) be amended to provide that:

Nothing in the agreement shall be construed to prevent the adoption or enforcement by any Contracting Party of measures necessary to protect human, animal or plant life or health, provided that these measures are consistent with sound scientific evidence and recognize the principle of equivalency.

To elaborate on the above amendment, GATT instruments should be drafted to provide that:

> The appropriate standards or guidelines of the Codex Alimentarius Commission, the International Office of Epizootics, the International Plant Protection Convention or, as appropriate, other scientific organizations open to full participation by all Contracting Parties, *e.g.*, the World Health Organization for situations involving hazards to human health and the environment, shall be considered by a panel in determining whether a measure designed to provide an acceptable level of protection is consistent with sound scientific evidence.

> A measure shall be deemed to be based on sound scientific evidence if the measure is equivalent to the appropriate standard established by an organization included above or if the measure was developed using information and analysis comparable to that used by such an organization. However, if there is not an international standard or guideline, or if a Party maintains a measure which is not equivalent to or has not been developed using information comparable to that used in an international standard or guideline, then Parties may still avail themselves of the dispute settlement procedures under the agreement.

> Measures which are not identical but which have the same effect in ensuring an acceptable level of protection shall be deemed to be equivalent.

In addition, GATT instruments should be drafted to provide that a notification, consultation and dispute settlement system having the following elements is available:

Notification: Each Contracting Party shall notify the GATT Secretariat of any proposed sanitary and phytosanitary regulation involving processes and production methods, product specifications and inspection and certification systems, as well as concluded bilateral agreements, which could have a significant effect on the trade of other Contracting Parties, it being understood that such notification would of itself be without prejudice to views on the consistency of measures with, or their relevance to, rights and obligations under the General Agreement.

Notification shall cover any technical regulations, standards, bilateral agreements or certification systems which have been adopted or proposed by central government bodies, by nongovernmental bodies which have legal power to enforce a technical regulation, or by regional standardizing bodies in which relevant bodies within parties' territories are members or participants.

The GATT Secretariat will, when it receives a sanitary or phytosanitary notification, circulate copies to all Contracting Parties and all interested international standardizing and certification bodies and draw the attention of developing country Contracting Parties to any notification relating to products of particular interest to them.

The normal time limit for comments on notifications shall be sixty days. Contracting Parties shall discuss comments upon request and take these comments and the results of these discussions into account.

Each Contracting Party shall ensure that an inquiry point exists through which sanitary and phytosanitary notifications can be forwarded to the GATT Secretariat, copies of all final regulations can be obtained, and all relevant inquiries can be directed.

Informal Consultations: Contracting Parties shall respond to requests for consultations promptly and attempt to conclude consultations expeditiously with a view to reaching a mutually satisfactory conclusion.

If a dispute is not resolved by consultations, the Contracting Parties involved in a dispute may request an appropriate body or individual to use their good offices with a view to the settlement of the outstanding differences between the Parties. The Contracting Parties are particularly encouraged to use the good offices of the international scientific organizations established to address sanitary and phytosanitary measures, *i.e.*, the Codex Alimentarius Commission, the International Office of Epizootics and the International Plant Protection Convention.

Dispute Settlement: Provisions regarding dispute settlement should be considered in consultation with the Negotiation Group on Dispute Settlement.

Existing National Approval Process: Some Contracting Parties maintain a domestic regime which generally requires the certification or approval of a broad class of products (*e.g.*, pharmaceuticals or pesticides) which may affect human, animal or plant life or health prior to the use or sale for use of those products within its territory. Before any other Contracting Party may initiate dispute settlement proceedings under this instrument, it shall have attempted to obtain certification or approval of the product in question in accordance with the rules of the regime, *provided* that the regime is intended to address a class of products which includes the product in question, that the regime uses reasonable and scientifically-based procedures and evidentiary standards to evaluate such products and that the regime's treatment of foreign products is no less favorable than that accorded to like products of national origin.

Relationship to International Standards and Organizations: The Codex Alimentarius Commission, the International Office of Epizootics, the International Plant Protection Convention, or other appropriate international scientific organizations shall be asked to provide a list of individuals with technical expertise in various areas. Regarding the consistency of a measure with sound scientific evidence, dispute settlement panels shall give primary consideration to the technical judgment of a technical advisory group composed of individuals selected from the appropriate list, its composition subject to the consent of the interested Contracting Parties.

In food safety, the following standards of the Codex Alimentarius and the associated scientific information and analysis shall be deemed to be based on sound scientific evidence: acceptable levels for food additives, maximum residue limits for veterinary drugs, allowable levels of environmental contaminants, maximuim residue limits for pesticides, methods of analysis and sampling, and codes and guidelines of hygienic practice.

In the area of animal health, the risk assessment guidelines developed under the auspices of the International Office of Epizootics for the use of the parties shall be deemed to be based on sound scientific evidence.

In the area of plant health, the risk assessment guidelines developed under the auspices of the International Plant Protection Convention for the use of the parties shall be deemed to be based on sound scientific evidence.

For matters not covered by the aforementioned standards or guidelines on food safety, animal health and plant health, the appropriate standards or guidelines of other scientific organizations open to full participation by all Contracting Parties shall be deemed to be based on sound scientific evidence.

If there is not an appropriate international standard or guideline, or if a Contracting Party maintains a measure which is not equivalent to or has not been developed using information and analysis comparable to that used in an international standard or guideline, then a Contracting Party shall have the option of using other experts, evidence, organizations, or other relevant sources of scientific information to show that its measures are consistent with sound scientific evidence.

National Treatment: The products of the territory of any Contracting Party shall be accorded treatment no less favorable than that accorded to like products of national origin in respect to all sanitary or phytosanitary laws, regulations, requirements, measures or approvals for use.

Implementation
Conforming amendments to the GATT instruments should be fully in effect in 1991.

Technical Assistance
The strengthening of the GATT approach to sanitary and phytosanitary measures may pose particular difficulties for developing countries. The Contracting Parties should evaluate the probable effects on developing countries of the enhanced GATT sanitary and phytosanitary procedures. If warranted by the results of this evaluation, the appropriate international organizations, for example, the United Nations Food and Agriculture Organization, might be contacted for technical assistance. The assistance provided might focus on strengthening the regulatory mechanisms of developing countries, particularly with regard to food safety and plant health, and could also facilitate the establishment of inquiry points where needed.

SPECIAL AND DISTINCTIVE TREATMENT FOR DEVELOPING COUNTRIES

Meaningful agricultural trade reform requires the active participation of all Contracting Parties. The new GATT rules and disciplines proposed for import access, export competition, internal support and sanitary and phytosanitary measures should be applicable to all Contracting Parties.

Developing countries with relatively advanced economies and/or well-developed agricultural sectors would be expected to comply fully with the implementation mechanisms identified in other sections of this paper. However, the United States understands the distinctive needs of less developed countries, particularly with respect to infrastructure and the difficulties some may have in implementing the transition schedules proposed for internal support and import access. In order to determine such needs, criteria related to the level of agricultural and overall development would need to be taken into account. The degree to which any developing country departs from the implementation schedules outlined in other parts of this paper should be commensurate with that country's demonstrated need for exceptional treatment.

Less-developed countries would be allowed to maintain agreed upon final bound tariffs on agricultural products at moderate levels commensurate with a particular country's demonstrated need for such treatment. As the overall economic performance of the country improves, these tariff levels would be progressively lowered to final bound rate comparable to those in effect for other Contracting Parties. Less-developed countries would also be allowed to maintain certain subsidies for the purpose of long-term agricultural development, provided they agree to progressively reduce such subsidies as the performance of their agricultural sector improves or the performance of their overall economy improves.

For products of priority export interest to developing countries, the negotiations should seek to provide reductions in trade barriers and internal support policies by developed countries on an accelerated basis.

ANNEX 1

Product Coverage

The United States proposes that the following products be subject to negotiations in the Negotiating Group on Agriculture. References are in accordance with the Harmonized System.

- Chapters 1 through 23 (agricultural and fisheries products)
- Heading 2401 (unmanufactured tobacco)
- Heading 3203 (coloring of vegetable or animal origin)
- Heading 3301 (essential oils)
- Headings 3501–3503 (casein, albumin and gelatin)
- Headings 4101–4103 (hides and skins other than furskins)
- Headings 4301–4302 (undressed and dressed furskins, not made into apparel)
- Headings 4401–4412 (wood and selected wood products)
- Headings 5101–5103 (wool and animal hair, not carded or combed, waste)
- Heading 5201–5202 (raw cotton and waste)

ANNEX 2

Illustrative List of Prohibited Export Subsidies

(a) The provision by governments of direct subsidies to a firm or an industry contingent upon export performance.

(b) Currency retention schemes or any similar practices which involve a bonus on exports.

(c) Internal transport and freight charges on export shipments, provided or mandated by governments, on terms more favorable than for domestic shipments.

(d) The delivery by governments or their agencies of imported or domestic products or services for use in the production of exported goods, on terms or conditions more favorable than for delivery of like or directly competitive products or services for use in the production of goods for domestic consumption, if (in the case of products) such terms or conditions are more favorable than those commercially available on world markets to their exporters.

(e) The full or partial exemption, remission, or deferral specifically related to exports, of direct taxes or social welfare charges paid or payable by industrial or commercial enterprises.

(f) The allowance of special deductions directly related to exports or export performance, over and above those granted in respect to production for domestic consumption, in the calculation of the base on which direct taxes are charged.

(g) The exemption or remission in respect of the production and distribution of exported products, of indirect taxes in excess of those levied in respect of the production and distribution of like products when sold for domestic consumption.

(h) The exemption, remission or deferral of prior stage cumulative indirect taxes on goods or services used in the production of exported products in excess of the exemption, remission or deferral of like prior stage cumulative indirect taxes on goods or services used in the production of like products when sold for domestic consumption; provided, however, that prior stage cumulative indirect taxes may be exempted, remitted or deferred on like products when sold for domestic consumption, if the prior stage cumulative indirect taxes are levied on goods that are physically incorporated (making normal allowance for waste) in the exported product.

(i) The remission or drawback of import charges in excess of those levied on imported goods that are physically incorporated (making normal allowance for waste) in the exported product; provided, however, that in particular cases a firm may use a quantity of home market goods equal to, and having the same quality and characteristics as, the imported goods as a substitute for them in order to benefit from this provision if the import and the corresponding export operations both occur within a reasonable time period, normally not to exceed two years.

(j) The provision by governments (or special institutions controlled by governments) of export credit guarantee or insurance programs, of insurance or guarantee programs against increases in the cost of exported products or of exchange risk programs, at premium rates, which are manifestly inadequate to cover the long-term operating costs and losses of the programs.

(k) The grant by governments (or special institutions controlled by and/or acting under the authority of governments) of export credits, at rates below those which they actually have to pay for the funds so employed (or would have to pay if they borrowed on international capital markets in order to obtain export credit), or the payment by them of all or part of the costs incurred by exporters or financial institutions in obtaining credits, insofar as they are used to secure a material advantage in the field of export credit terms.

Provided, however, that if a signatory is a party to an international undertaking on official export credits to which at least twelve original signatories to the Agreement are parties as of January 1, 1979 (or a successor undertaking which has been adopted by those original signatories), or if in practice a signatory applies the interest rates provisions of the relevant undertaking, an export credit practice which is in conformity with those provisions shall not be considered an export subsidy prohibited by this Agreement.

(1) Any other charge on the public account constituting an export subsidy in the sense of Article XVI of the General Agreement.

ANNEX 3

INTERNAL POLICY CATEGORIES

Policies to be Phased Out

Administered price policies, including administered prices (loan rates, intervention prices, others) resulting from dual pricing policies, state control, marketing boards, domestic price controls and consumer levies; certain subsidies to processors; etc.

Income support policies, including income- and price-stabilization payments keyed to production, price or cost of production, marketing loans; payment to convert production from one commodity to another; headage payments, etc.

Marketing subsidies, including transportation subsidies provided only to agricultural outputs, programs that provide price discounts to purchasers, etc.

Input subsidies that are not provided to producers or processors of agricultural commodities on an equal basis, including subsidies on fertilizer, pesticides, water from irrigation projects, production credit, subsidized raw materials (feed ingredients or raw materials for processed products), fuel or electricity subsidies, etc.

Investment subsidies that are not provided to producers or processors of agricultural commodities on an equal basis, including government provision of subsidized capital, long-term loans, breeding stock or perennial stock, farm modernization or consolidation programs, etc.

Permitted Policies

Direct income payments to producers or processors that are not linked to current production, price, cost of production, or marketing of agricultural commodities, including flat-rate income transfers not linked to current production, etc.

Environmental and conservation programs, including funding to assist the adoption of *bona fide* conservation practices, etc.

Disaster assistance keyed to *bona fide* production losses, including disaster payments, crop insurance, disaster relief, etc.

Domestic food aid based on need, including food donations, food stamps, programs for particular consumer groups, etc.

Marketing programs that do not confer an economic benefit—in the form of price discounts, cash or in-kind payments including market development programs meeting this criteria, market information programs, inspection and grading programs, etc.

General services, including government-funded research, extension, pest and disease control, education programs, etc.

Resource retirement programs, including direct payments to remove land or other production factors from agriculture, retraining programs, early retirement schemes, etc.

Programs for stockpiling food reserves that do not provide direct price or income support or subsidized inputs to producers, processors or consumers.

Policies to be Disciplined

All other agricultural programs not elsewhere specified, including, but not limited to:

- certain policies that do not meet criteria developed for permitted programs or policies to be phased out;
- input subsidies provided to any producer or processor of agricultural commodities on an equal basis, including subsidies on fertilizer, pesticide, water from irrigation projects, production credit, subsidized raw materials (feed ingredients or raw materials for processed products, fuel or electricity subsidies, etc.;
- investment subsidies provided to any producer or processor of agricultural commodities on an equal basis, including government provision of subsidized capital, long-term loans, breeding stock or perennial stock, farm modernization or consolidation programs, etc.

Appendix B
The Effects of Tariffs and Import Quotas

Economic theory provides comparisons between the effect of tariffs and the effects of import quotas.[1] The tendency on the part of both is to raise internal prices, reduce overall consumption, limit imports and increase domestic production. Each measure, however, affects revenue differently. Governments collect duties resulting from tariffs, but under a quota policy, the organizational structure of importers and exporters will dictate who is to receive the "rent" resulting from increased domestic prices. Quantitative import controls and tariffs located for various political and economic reasons among competing countries, often in a discriminatory manner and frequently without regard to cost-price conditions. Tariffs do not discriminate among exporters enjoying most-favored-nation status. In certain circumstances, quantitative restrictions, including import quotas, may be combined with specified minimum rates of duty to form "tariff quotas," with additional imports being admitted without limitation subject to payment of higher rates of duty.[2]

Economists have endeavored to analyze the incidence and economic effects of import quotas and attendant devices, often forgetting a singularly important characteristic, that of administrative flexibility. Tariffs are, as a rule, stable over time and tariff legislation—even that relating to "peril points," "escape clauses" and the like—takes time to change. This is not the case with import quotas. They are temporary by nature and subject to administrative discretion.

Tariffs are usually imposed at a sensitive point of the price mechanism and their effects are spread among the variables affecting supply and demand. They can be imposed at a sufficiently high level to keep a commodity out of a country, although circumstances may prevail which

161

allow imports to continue even with high tariffs—for example, when the effect of the duty is borne by the foreign exporter or can be passed on by an intermediate user.

The situation of exporting countries with highly inelastic supply curves is the most important case in quota theory. In these circumstances, a tariff can neither materially increase the price in the importing country nor reduce the volume of imports. While the tariff improves the terms of trade and the importing government gains revenue by taxing the foreigner, political circumstances among producers at home often dictate a different "solution." Quotas have provided that solution.

The case described above is illustrated in Figure B.1. The system is assumed to be closed. Total supply and demand conditions for a particular commodity are represented in the figure by the importing country (Country M) and exporting country (Country X). In this situation, no reasonable tariff will raise and redistribute incomes among producers in Country M. On the other hand, a quota which halves imports would raise prices from OP to OP$'$. If Country X did nothing about this quota, the price there would fall to a level approaching P$''$. Hence the imposition of the quota by Country M is likely to be followed by a stabilization scheme in Country X .[3]

Notes and Sources

1.   For an excellent treatment of tariff and quota theory see W.M. Corden, *The Theory of Protection* (Oxford: Clarendon, 1971).

2.   For a detailed discussion of tariff quotas, see Michael Rom, *Role of Tariff Quotas in Commercial Policy* (London: Macmillan for the Trade Policy Research Centre, 1979).

3.   Charles P. Kindleberger, *International Economics*, Fourth edition (Homewood, Illinois: Richard D. Irwin , Inc., 1968), pp. 132–4.

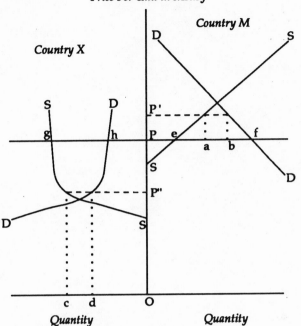

FIGURE B.1. The Imposition of Quotas in Partial Equilibrium in Face of
Inelastic Supply.

Note: ab = cd - imports equal exports at prices P' and P"
         ef = gh - the shortfall in Country M is equal to the surplus in
                     Country X and price P (world price)
         OP" < OP'   the price in Country X is lower than the price of the same
                      commodity in Country M after the imposition of a quota by
                      Country M; thus prices and values are different.

Source: Charles P. Kindleberger, *International Economics* , Fourth edition
(Homewood, Illinois: Richard D. Irwin, Inc., 1968), 133.

Appendix C

Section 22 of the Agricultural Adjustment Act of 1933, as Reenacted and Amended[1]

(a) Whenever the Secretary of Agriculture has reason to believe that any articles are being or are practically certain to be imported into the United States under such conditions and in such quantities as to render or tend to render ineffective, or materially interfere with, any program or operation undertaken under this title or the Soil Conservation and Domestic Allotment Act, as amended, or Section 32, Public Law Numbered 320, Seventy-fourth Congress, approved August 24, 1935, as amended, or any loan, purchase, or other program or operation undertaken by the Department of Agriculture, or any agency operating under its direction, with respect to any agricultural commodity or product thereof, or to reduce substantially the amount of any product processed in the United States for any agricultural commodity or product thereof with respect to which any such program or operation is being undertaken, he shall so advise the President, and, if the President agrees that there is reason for such belief, the President shall cause an immediate investigation to be made by the United States Tariff Commission,[2] which shall be given precedence to investigations under this section to determine such facts. Such investigation shall be made after due notice and opportunity for hearing to interested parties, and shall be conducted subject to such regulations as the President shall specify. (7 USC 624(a)).

(b) If, on the basis of such investigation and report to him of findings and recommendations made in connection therewith, the President finds the existence of such facts, he shall by proclamation impose such fees not in excess of 50 per centum *ad valorem* or such quantitative limitations on any article or articles which may be entered, or withdrawn from warehouse, for consumption as he finds and declares shown by such investigation to be necessary in order that the entry of such article or articles will not render or tend to render ineffective, or materially interfere with, any program or operation referred to in subsection (a) of this section, or reduce substantially the amount of any product processed in the United States from any such agricultural commodity or product thereof with respect to which any such program or operation is being undertaken: *Provided*, that no proclamation under this section shall impose any limitation on the total quantity of any article or articles which may be entered, or withdrawn from warehouse, for consumption which reduces such permissible total quantity to proportionately less than 50 per centum of the total quantity of such article or articles which was entered, or withdrawn from warehouse, for consumption during a representative period as determined by the President: *And provided further*, that in designating any article or articles, the President may describe them by physical qualities, value, use or upon such other bases as he shall determine.

In any case where the Secretary of Agriculture determines and reports to the President with regard to any article or articles that a condition exists requiring emergency treatment, the President may take immediate action under this section, without awaiting the recommendations of the Tariff Commission, such action to continue in effect pending the report and recommendations of the Tariff Commission and action thereon by the President. (7 USC 644(b).)[3]

(c) The fees and limitations imposed by the President by proclamation under this section and any revocation, suspension, or modification thereof, shall become effective on such date as shall be therein specified and such fees shall be treated for administrative purposes and for the purposes of Section 32 of Public Law Numbered 320, Seventy-fourth Congress, approved August 24, 1935, as amended, as duties imposed by the Tariff of 1930, but such fees shall not be considered as duties for the purpose of granting any preferential concession under any international obligation of the United States. (7 USC 624(c).)

(d) After investigation, report, findings, and declaration in the manner provided in the case of a proclamation issued pursuant to subsection (b) of this section, any proclamation or provision of such proclamation may be suspended or terminated by the President whenever he finds and proclaims that the circumstances requiring the proclamation or provision thereof no longer exist or may be modified by the President whenever he finds and proclaims that changed circumstances required such modification to carry out the purposes of this section. (7 USC 624(d).)

(e) Any decision of the President as to facts under this section shall be final. (7 USC 624(e).)

(f) No trade agreement or other international agreement heretofore or hereafter entered into the United States shall be applied in a manner inconsistent with the requirements of this section. (7 USC 624(f).)[4]

Notes and Sources

1. See also section 202(a) of the Agricultural Act of 1956. Section 22 was added by the Act of August 24, 1935 (49 Stat. 773). As originally enacted, action under this section could be taken only with respect to articles the importation of which was found to be adversely affecting programs or operations under the Agricultural Adjustment Act of 1933. Section 22 has been amended several times and was revised in its entirety by Section 3 of the Agricultural Act of 1948 (62 Stat. 1247) and again by Section 3 of the Act of June 28, 1950 (64 Stat. 261). Regulations governing investigations under this section are set forth in Excecutive Order 7233, dated November 23, 1935, and in 19 CFR 201, 204.

2. Name changed to the United States International Trade Commission by the Trade Act of 1974 (Public Law 93–618).

3. Paragraph added by Section 104 of the Trade Agreements Extension Act of 1953, 67 Stat. 472.

4. The provisions of this subsection (f) were substituted for earlier provisions of section 8(b) of the Trade Agreements Extension Act of 1951, approved June 16, 1951, 65 Stat. 72, 75.

Appendix D
Meat Import Act of 1979

Be it enacted by the Senate and House of Representatives of the United States of America in Congress assembled. That section 2 of the Act of August 22, 1964, entitled "An Act to provide for the free importation of certain wild animals, and to provide for the imposition of quotas on certain meat and meat products" (19 USC 1202 note) is amended to read as follows:

Section 2

(a) This section may be cited as the "Meat Import Act of 1979."

(b) For purposes of this section—

    (1) The term "entered" means entered, or withdrawn from warehouse, for consumption in the customs territory of the United States.

    (2) The term "meat articles" means the articles provided for in the Tariff Schedules of the United States (19 USC 1202) under—

        (A) item 106.10 (relating to fresh, chilled or frozen meat);

        (B) items 106.22 and 106.25 (relating to fresh, chilled, or frozen meat of goats and sheep [except lambs]); and

        (C) items 107.55 and 107.62 (relating to prepared and preserved beef and veal [except sausage]) if the articles are prepared, whether fresh, chilled, or frozen, but not otherwise preserved.

(3) The term "Secretary" means the Secretary of Agriculture.

(c) The aggregate quantity of meat articles which may be entered in any calendar year after 1979 may not exceed 1,204,600,000 pounds; except that this aggregate quantity shall be—

   (1) increased or decreased for any calendar year by the same percentage that the estimated average annual domestic commercial production of meat articles in that calendar year and the two preceding calendar years increases or decreases in comparison with the average annual domestic commercial production of meat articles during calendar years 1968 through 1977; and

   (2) adjusted further under subsection (d).

       For purposes of paragraph (1), the estimated annual domestic commercial production of meat articles for any calendar year does not include the carcass weight of live cattle specified in items 100.40, 100.43, 100.45, 100.53, and 100.55 of such Schedules entered during such year.

(d) The aggregate quantity referred to in subsection (c), as increased or decreased under paragraph (1) of such subsection, shall be adjusted further for any calendar year after 1979 by multiplying such quantity by a fraction—

   (1) the numerator of which is the average annual per capita production of domestic cow beef during that calendar year (as estimated) and the four calendar years preceding such calendar year; and

   (2) the denominator of which is the average annual per capita production of domestic cow beef in that calendar year (as estimated) and the preceding calendar year.

       For the purposes of this subsection, the phrase "domestic cow beef" means that portion of the total domestic cattle slaughter designated by the Secretary as cow slaughter.

(e) For each calendar year after 1979, the Secretary shall estimate and publish—

(1) before the first day of such calendar year, the aggregate quantity prescribed for such calendar year under subsection (c) as adjusted under subsection (d); and

(2) before the first day of each calendar quarter in such calendar year, the aggregate quantity of meat articles which (but for this section) would be entered during such calendar year.

In applying paragraph (2) for the second or any succeeding calendar quarter in any calendar year, actual entries for the preceding calendar quarter or quarters in such calendar year shall be taken into account to the extent data are available.

(f) (1) If the aggregate quantity estimated before any calendar quarter by the Secretary under subsection (e)(2) is 110 percent or more of the aggregate quantity estimated by him under subsection (e)(1), and if there is no limitation in effect under this section for such calendar year with respect to meat articles, the President shall by proclamation limit the total quantity of meat articles which may be entered during such calendar year to the aggregate quantity estimated for such calendar year by the Secretary under subsection (e)(1); except that no limitation imposed under this paragraph for any calendar year may be less than 1,250,000,000 pounds. The President shall include in the articles subject to any limit proclaimed under this paragraph any article of meat provided for in item 107.61 of the Tariff Schedules of the United States (relating to high-quality beef specially processed into fancy cuts).

(2) If the aggregate quantity estimated before any calendar quarter by the Secretary under subsection (e)(2) is less than 110 percent of the aggregate quantity estimated by him under subsection (e)(1), and if a limitation is in effect with respect to meat articles, such limitation shall cease to apply as of the first day of such calendar quarter. If any such limitation has been in effect for the third calendar quarter of any calendar year, then it shall continue in effect for the fourth calendar quarter of such year unless the proclamation is suspended or the total quantity is increased pursuant to subsection (g).

(g) The President may, after providing opportunity for public comment by giving 30 days' notice by publication in the Federal Register of his intention to so act, suspend any proclamation made under subsection (f), or increase the total quantity proclaimed under such subsection, if he determines and proclaims that—

(1) such action is required by overriding economic or national security interests of the United States, giving special weight to the importance to the nation of the economic well–being of the domestic cattle industry;

(2) the supply of meat articles will be inadequate to meet domestic demand at reasonable prices; or

(3) trade agreements entered into after the date of enactment of this Act insure that the policy set forth in subsections (c) and (d) will be carried out.

Any such suspension shall be for such periods, and any such increase shall be in such amount, as the President determines and proclaims to be necessary to carry out the purposes of this subsection.

(h) Notwithstanding the previous subsections, the total quantity of meat articles which may be entered during any calendar year may not be increased by the President if the fraction described in subsection (d) for that calendar year yields a quotient of less than 1.0, unless—

(1) during a period of national emergency declared under section 201 of the National Emergencies Act of 1976, he determines and proclaims that such action is required by overriding national security interests of the United States;

(2) he determines and proclaims that the supply of articles of the kind to which the limitation would otherwise apply will be inadequate, because of a natural disaster, disease or major market disruption, to meet domestic demand at reasonable prices, or

(3) on the basis of actual data for the first two quarters of the calendar year, a revised calculation of the fraction described in subsection (d) for the calendar year yields a quotient of 1.0 or more.

Any such suspension shall be for such period, and any such increase shall be in such amount, as the President determines and proclaims to be necessary to carry out the purposes of this subsection. The effective period of any such suspension or increase made pursuant to paragraph (1) may not extend beyond the termination, in accordance with the provisions of section 202 of the National Emergencies Act of 1976, of such period of national emergency, notwithstanding the provisions of section 202(a) of that Act.

(i) The Secretary shall allocate the total quantity proclaimed under subsection (f)(1) and any increase in such quantity provided for under subsection (g) among supplying countries on the basis of the shares of the United States market for meat articles such countries supplied during a representative period. Notwithstanding the preceding sentence, due account may be given to special factors which have affected or may affect the trade in meat articles or cattle. The Secretary shall certify such allocations to the Secretary of the Treasury.

(j) The Secretary shall issue such regulations as he determines to be necessary to prevent circumvention of the purposes of this section.

(k) All determinations by the President and the Secretary under this section shall be final.

(1) The Secretary of Agriculture shall study the regional economic impact of imports of meat articles and report the results of his study, together with any recommendations (including recommendations for legislation, if any) to the Committee on Ways and Means of the House of Representatives and to the Committee on Finance of the Senate not later than June 30, 1980.

SEC. 2. This Act shall take effect January 1, 1980.

Approved December 31, 1979.

Appendix E
Countervailing Duty versus Nontariff Measures[1]

The economic arguments for a countervailing duty against subsized exports are illustrated in Figure E.1. We will assume that the importing country is unable to affect the world price of the imported good. Under free trade and no export subsidy from world suppliers, the domestic market is in equilibrium at Point A, where the supply curve (S_f) intersects the demand curve (D). This point maximizes world gains from trade since the marginal value of an extra unit, and the marginal cost of supplying an additional unit are equal at P_0.

An export subsidy of P_0P_1 lowers the supply curve to Sf'; the market falls to P_1 and imports increase to Q_1 from Q_0. From a global view, this policy is inefficient because excess resources are used in production. The importing country is encouraged to consume Q_1 of the good when price is P_1, yet world costs are P_0. Area ACP_1P_0 represents the importing country's net gain, while area CBP_0P_1 represents the total export subsidy bill. Thus, the net world resource loss is equivalent to the area ABC.

If the importing country applied a countervailing duty just large enough to offset the export subsidy, trade would return to the original price (P_0) and volume (Q_0) as with free trade and no subsidy (Point A). In terms of world efficiency, area ABC, representing wasted resources, is eliminated.

The countervailing duty in this case is an application of the "specification rule." The use of policy tools closest to the locus of the distortions separating private and social costs is most efficient. When excess exports are exactly matched by a countervailing duty equaling the value of the distorting subsidy, trade volume is unaffected. In this case, however, the exporting country provides a financial transfer to the

176

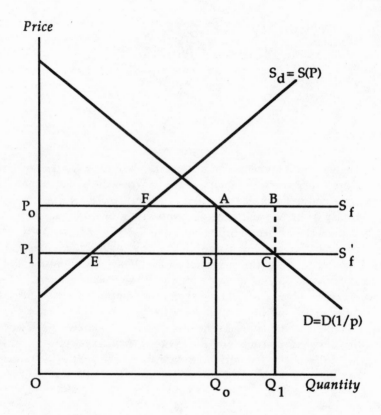

FIGURE E.1. Export Subsidy and Countervailing Duty.

importing government equal to area ADEF. Thus, export subsidies do nothing for domestic producers in the presence of countervailing duties; all the economic gains are captured by the government of the importing country.

Nontariff trade barriers can affect the terms and volume of trade in the same manner as a countervailing duty. Nontariff barriers cause the price of the import-competing good to increase as the quantity of the import supplied is reduced. This is generally done by limiting the volume of imports permitted—by imposing significant costs on foreign producers, exporters or domestic importers; or by imposing conditions of uncertainty on importers to which they respond by curtailing imports. If the importing country introduces a trade measure that increases costs of production in the exporting country to the original level (P_0), imports will also return to Q_0. The difficulty, however, lies in imposing a nontariff measure which raises costs in an amount exactly equal to the export subsidy. Although an equivalent tariff value can be computed for a nontariff measure *ex post*, it is difficult to estimate the welfare effects of any such measure *ex ante*. Nontariff measures are unlikely, therefore, to be an efficient policy tool to offset distortions.

A diagram of the import-directed trade distortion is provided in Figure E.2. Foreign supply is assumed perfectly elastic at price P_1. At P_1 imports are Q_1Q_4 and domestic output is OQ_1. To reduce the competition from imports without imposing a tariff, the country can impose a nontariff measure with an equivalent price effect of P_0P_1. Import volume is reduced to Q_2Q_3, and the supply curve becomes $S+C$ reflecting the additional cost incurred by exporters. Total loss of consumer surplus is P_1BAP_0, of this P_1EFP_0 is gained directly by domestic producers because of price increases. The domestic economy incurs real losses in the form of consumption and production effects, areas ABC and DEF, respectively.

If the nontariff measure takes the form of a quota, the area ACDF represents profit to importers or foreign exporters. If safety standards or other requirements are applied equally to imports and to import-competing goods there is no change in trade, since the foreign and the domestic supply functions shift by an equal amount. In this case, there is no welfare redistribution or financial transfer between producers. There is, however, a consumption effect because of the higher-priced imported good.

On the other hand, if the cost-increasing requirements are applied only to imports, the area ACDF may be lost in the same manner as area ABC and DEF under a quota. The real cost of the nontariff measure then is area P_0ABP_1 or the additional cost of processing incurred by foreign producers.

178

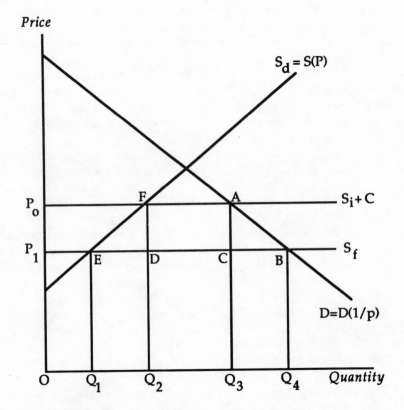

FIGURE E.2. Cost of Nontariff Trade Barriers.

Notes and Sources

1.   Material   in this Appendix is adapted from Mark B. Lynham, *Nontariff Agricultural Trade Barriers: Livestock and Meat Legislative Regulation Devices as they Affect International Trade Between Industrially Developed Countries.*   Unpublished Master's Thesis, Department of Agricultural Economics, University of Arizona, 1983.

Bibliography

Ackerman, Karen Z., and Mark E. Smith. *Agricultural Export Programs: Background for* 1990 *Farm Legislation,* No. AGES9033. Washington, D.C.: Commodity Economics Division, Economic Research Service, USDA, May 1990.

Aho, C. Michael, and Jonathan David Aronson. *Trade Talks: America Better Listen!* New York: Council on Foreign Relations, 1985.

Anderson, Kym, and Yujiro Hayami. *The Political Economy of Agricultural Protection: East Asia in International Perspective.* Sydney: Allen & Unwin, 1986.

Annual Report of the Council on International Economic Policy. Washington, D.C.: U.S. Government Printing Office, February, 1974.

Atlantic Council. *U.S. Agriculture in a World Context: Policies and Approaches for the Next Decade.* Washington, D.C.: Praeger, 1973.

Auge-Laribe, M. *La Politique agricole de la France de* 1880 *à* 1940. Paris: Presses Universitaires de France, 1950.

Balassa, B. *The Structure of Protection in Developing Countries.* Baltimore: Johns Hopkins University Press, 1971.

Baldwin, Robert D. *Nontariff Distortions of International Trade.* Washington, D.C.: Brookings Institution, 1970.

Bane, F. "Administrative Marketing Barriers." *Law and Contemporary Problems* 8, No. 2 (Spring 1941): 376–90.

Benedict, M.R. *Farm Policies of the United States,* 1790–1950. New York: Twentieth Century Fund, 1953.

Bhagwati, J., et al. *Trade Balance of Payments and Growth.* Amsterdam: North Holland, 1971.

181

Bidwell, Percy. *The Invisible Tariff.* New York: Council on Foreign Relations, 1939.

Boardman, Robert. *Pesticides in World Agriculture: The Politics of International Regulation.* London: Macmillan, 1986.

Britton, D.K. "Problems of Adapting U.K. Institutions in the EEC." Conference Paper No. 9, Agricultural Adjustment Conference, University of Newcastle-upon-Tyne, 1972.

Bureau of Agricultural Economics. *Agricultural Policies in the European Community: Their Origins, Nature and Effects on Production and Trade.* Policy Monograph No. 2. Canberra: Australian Government Publishing Service, 1985.

Commission on International Trade and Investment Policy. *United States International Economic Policy in an Interdependent World.* Vols. 1 and 2. Washington, D.C.: U.S. Government Printing Office, July 1971.

Condliffe, J.B. *The Commerce of Nations.* London: George Allen and Unwin, 1951.

———. *The Reconstruction of World Trade.* New York: Norton, 1940.

Congressional Record. Washington, D.C.: U.S. Government Printing Office, June 1968.

Corden, W.M. "Tariffs and Protectionism." In *The International Encyclopedia of Social Sciences,* Vol. 8. New York: Macmillan Co. and Free Press, 1968, pp. 113–21.

———. *The Theory of Protection.* Oxford: Clarendon Press, 1971.

———. *Trade Policy and Economic Welfare.* London: Oxford University Press, 1974.

Curzon, Gerard and Victoria. *Global Assault on Nontariff Trade Barriers.* Thames Essay No. 3. London: Trade Policy Research Centre, 1972.

———. *Hidden Barriers to International Trade.* Thames Essay No. 1. London: Trade Policy Research Centre, 1970.

Deardorff, Alan V., and Robert M. Stern. *Methods of Measurement of Nontariff Barriers.* United Nations Conference on Trade and Development, UNCTAD/ST/MD/28, January 2, 1985.

Denton, Geoffrey, and Seamus O'Cleireacain. *Subsidy Issues in International Commerce.* Thames Essay No. 5. London: Trade Policy Research Centre, 1972.

EEC Commission. *Fourth General Report on the Activities of the Communities,* 1970. Brussels: EEC, February 1971.

Ehrich, Rollo L., and Mohammad Usman. *Demand and Supply Functions for Beef Imports.* Bulletin 604. Laramie: Agricultural Experiment Station, University of Wyoming, January 1974.

FAO. *Developments in Agricultural Price Stabilization and Support Policies*, 1965–1970. Committee on Commodity Problems, CCP 71/15. Rome, August 1971.

———. *International Agricultural Adjustment: A Case Study of Japan.* C/73/LIM/3. Rome, 1973.

———. *National Regulations Affecting Imports of Livestock and Meat.* CCP:ME 72–8. Rome, April 1972.

———. *Note on Tariff and Nontariff Measures Affecting International Trade in Meat and Poultry.* Committee on Commodity Problems, CCP: Mah 70/b. Rome, May 1970.

———. *Processed Agricultural Products and Agricultural Adjustment.* C73/LIM/11. Rome, November 1973.

GATT. *International Trade*, 1975–76. Geneva: GATT, 1976.

———. *Trade Policies for a Better Future: Proposals for Action.* Geneva: GATT, March 1985.

Golt, Sidney. *The GATT Negotiations, 1973–75: A Guide to the Issues.* London: British-North American Committee, April 1974.

Gorman, William D., Gail A. Welsh, and Jimmye S. Hillman, eds. *Research Opportunities in Beef Export Markets: United States and Pacific Rim Countries.* Proceedings of an International Symposium by W-177 Western Regional Beef Research Project and Farm Foundation, at University of Arizona, Tucson, November 19–20, 1986. Report No. 621, Western Regional Research, June 1987.

Great Britain Ministry of Agriculture, Fisheries and Food. *Annual Review and Determination of Guarantees.* London: HM Stationery Office, 1969, 1971.

Haberler, G., *et al.* *Trends in International Trade.* Geneva: GATT, 1958.

Hathaway, Dale E. *Agriculture and the GATT: Rewriting the Rules.* Policy Analyses in International Economics 20, Washington, D.C.: Institute for International Economics, September 1987.

Hill, B.E. "The World Market for Beef and Other Meat." In *World Animal Review*, No. 4, Rome, 1972.

Hillman, J.S. *Economic Aspects of Interstate Agricultural Trade Barriers in the Western Region.* Ph.D. dissertation, University of California, Berkeley, 1954.

Hillman, J.S., and Robert A. Rothenberg. *Agricultural Trade and Protection in Japan.* Thames Essay No. 52. Aldershot, England: Gower, for the Trade Policy Research Centre, 1988.

Hillman, J.S., and J.D. Rowell. *A Summary of Laws Relating to the Interstate Movement of Agricultural Products in the Eleven Western States.* Tucson: University of Arizona Agricultural Experiment Station, 1952.

Hindley, Brian. *Britain's Position on Nontariff Protection.* London: Trade Policy Research Centre, 1972.

——. "Negotiations on Nontariff Barriers." Chapter 6 in *Towards an Open World Economy.* London: Macmillan, 1972.

Hoover, Edgar M. *The Location of Economic Activity.* New York: McGraw-Hill Book Co., 1948.

Hudec, Robert E. *Adjudication of International Trade Disputes.* Thames Essay No. 16. London: Trade Policy Research Centre, 1978.

——. *Developing Countries in the GATT Legal System.* Thames Essay No. 50. Aldershot, England: Gower, for the Trade Policy Research Centre, 1987.

Institute for International Economics and Institute for Research on Public Policy. *Reforming World Agricultural Trade: A Policy Statement by Twenty-nine Professionals from Seventeen Countries.* Washington, D.C.: Institute for International Economics, May 1988.

International Trade Center. *The Market for Manufacturing Grade Beef in the United Kingdom and the European Economic Community.* Geneva: UNCTAD, GATT, 1971.

Jasiorowsky, H.A. "Twenty Years with No Progress?" In *World Animal Review,* No. 4, Rome, 1972.

Johnson, D. Gale. "Are High Farm Prices Here to Stay?" *The Morgan Guaranty Survey.* New York: Morgan Guaranty Trust Co., August 1974.

——. *World Agriculture in Disarray.* London: Fontana-Collins, 1973.

Johnson, D. Gale, Kenzo Hemmi, and Pierre Lardinois. *Agricultural Policy and Trade: Adjusting Domestic Programs in an International Framework.* Triangle Papers 29. New York: New York University Press, 1985.

Johnson, H.G. "An Economic Theory of Protectionism, Tariff Bargaining and the Formation of Customs Unions." *Journal of Political Economy* 73 (June 1965):254–83.

Josling, T. *Agriculture and Britain's Trade Policy Dilemma.* Thames Essay No. 2. London: Trade Policy Research Centre, 1970.

——. *Agriculture in the Tokyo Round Negotiations.* Thames Essay No. 10. London: Trade Policy Research Centre, 1977.

——. "The EEC and the World Market for Temperate Zone Agricultural Products." Paper presented at the Royal Economic Conference, York, England, September 25–28, 1972.

Josling, T., and J.S. Hillman. *Agricultural Protection and Stabilization Policies: A Framework of Measurement in the Context of Agricultural Adjustment.* FAO C75/LIM/2. Rome: FAO, 1975.

Kelly, Margaret, *et al. Issues and Developments in International Trade Policy.* Occasional Paper No. 63. Washington, D.C.: International Monetary Fund, December 1988.

Langhammer, Rolf J., and André Sapir. *Economic Impact of Generalized Tariff Preferences.* Thames Essay No. 49. Aldershot, England: Gower, for the Trade Policy Research Centre, 1987.

Leipmann, H. *Tariff Levels and the Economic Unity of Europe.* London: George Allen and Unwin, 1938.

Long, Oliver, "The Multilateral Trade Negotiations in a Changing World Economic." Address to the Swedish National Committee of International Chambers of Commerce, Stockholm, March 19, 1974.

McArthur, M.H. "Australia's Role in the World's Beef Future." San Antonio: Paper presented at meeting of the American National Cattlemen's Association, January 1973.

McKnight, Frank G. *Foreign Agriculture.* Washington, D.C.: Foreign Agricultural Service, USDA, 1972.

Malmgren, H.B., "Negotiating Nontariff Barriers: The Harmonization of National Economic Policies." In *U.S. Foreign Economic Policy for the 1970s: A New Approach to New Realities.* Washington, D.C.: National Planning Association, 1971, pp. 79–109.

——. *Trade Wars or Trade Negotiations: Nontariff Barriers and Economic Peacemaking.* Washington, D.C.: Atlantic Council of the United States, 1970.

Malmgren, H.B., and D.L. Schlecty. "Rationalizing World Agricultural Trade." *Journal of World Trade Law* 4 (July-August, 1970):515–47.

Miller, Geoff. *The Political Economy of International Agricultural Policy Reform.* Canberra: Australian Government Publishing Service, 1986.

Mun, T. *England's Treasure by Forraign Trade.* London and Oxford: published for the Economic History Society by B. Blackwell, 1949.

National Center for Food and Agricultural Policy, Resources for the Future. *Mutual Disarmament in World Agriculture: A Declaration on Agricultural Trade.* Washington, D.C.: Resources for the Future, May 1988.

National Commission on Agricultural Trade and Export Policy. *Interim Report to the President and the Congress.* March 1985.

Needham, Peter, J. "How Ireland Conquered the British Beef Market." *Calf News* 12, No. 3 (March 1974):54, 55, 82.

New Zealand Meat Producers Board. "Annual Report and Statement of Account for Year Ended September 30, 1971." Wellington, 1972.

——. *New Zealand Meat Markets, 1970.* Wellington, March 19, 1971.

——. "New Zealand Meat Seminar." Papers presented in London, October 16–18, 1972.

OECD. *Policy Perspectives for International Trade and Economic Relations.* Paris: OECD, 1970.

Ohlin, G. "Trade in a Non-Laissez Faire World." In *International Economic Relations,* edited by Paul Samuelson. London: Macmillan, 1969, pp. 157-75.

Ojala, E.M. *Agricultural and Economic Progress.* London: Oxford University Press, 1952.

——. "Europe and the World Agricultural Economy." Conference paper No. 1. Agricultural Adjustment Conference, University of Newcastle-upon-Tyne, July 1972.

Patterson, Gardiner. *Commission on International Trade and Investment Policy.* Vol. 1. Washington, D.C.: U.S. Government Printing Office, 1971.

Power, A.P., and S.A. Harris. "A Cost/Benefit Evaluation of Alternative Policies for Foot-and-Mouth Disease in Great Britain." *Journal of Agricultural Economics* 24, No. 3 (1973):573-98.

Raney, Terri, and David Kelch. "The Safe Food Issue: New Nontariff Barriers?" *World Agriculture Situation and Outlook Report,* WAS–59. Washington, D.C.: Economic Research Service, USDA, June 1990, pp. 28-32.

Röpke, W. *German Commercial Policy.* London: Longmans, 1934.

Sanderson, Fred H. *Agriculture and International Trade.* Washington, D.C.: Council on U.S. International Trade Policy, February 1988.

Sanderson, Fred H., ed. *Agricultural Protectionism in the Industrialized World.* Washington, D.C.: Resources for the Future, 1990.

Schwinger, Robert B. "New Concepts and Methods in Foreign Trade Negotiations." *American Journal of Agricultural Economics* 51, No. 5 (December 1969):1338–49.

Stoeckel, Andy. *Intersectoral Effects of the CAP: Growth, Trade and Unemployment.* Bureau of Agricultural Economics, Occasional Paper No. 95. Canberra: Australian Government Publishing Service, 1985.

Study Group of the British Coordinating Committee for International Studies. "Memorandum on British External Economic Policy in Recent Years, Part II." 1939.

Tobin, J. "Economic Growth as an Objective of Government Policy." *American Economic Review* 54 (1964):1–200.

Tracy, M. *Agriculture in Western Europe: Challenge and Response,* 1880–1980. (London: Granada, 1982) pp. 25–130.

Trade Policy Research Centre. "British, European and American Interests in International Negotiations on Agricultural Trade." Staff Paper No. 3, London, 1973.

Tutwiler, M. Ann, and George E. Rossmiller. *Prescriptions for Success in the GATT.* Washington, D.C.: Resources for the Future, December 1987.

United Nations. Conference on Trade and Development. *Commodity Problems and Policies: Access to Markets. A Report by the UNCTAD Secretariat.* Santiago, Chile: UNCTAD, 1972.

——. TD120/Supp. 1, January 31, 1972.

United States. Congress. Congressional Budget Office. *The Gatt Negotiations and U.S. Trade Policy.* Washington, D.C.: U.S. Government Printing Office, June 1987.

United States. Department of Agriculture. *Code of Agriculture. Section 22 of the Agricultural Adjustment Act of 1933 as Reenacted and Amended.* Washington, D.C., January 1972.

——. *Measures of the Degree and Cost of Economic Protection of Agriculture in Selected Countries.* Washington, D.C., 1967.

United States. Department of Agriculture. Economic Research Service. *Prospects for Agricultural Trade with the USSR.* Foreign Series Report No. 356. Washington, D.C., 1973.

United States. Department of Agriculture. Economic Research Service. Agriculture and Trade Analysis Division. *Agriculture in the Uruguay Round: Analyses of Government Support.* Staff Report AGES880802. Washington, D.C.: U.S. Government Printing Office, December 1988.

——. *Bibliography of Research Supporting the Uruguay Round of the GATT.* Staff Report No. AGES 89-64. Washington, D.C.: U.S. Government Printing Office, December 1989.

United States. Department of Agriculture. Economic Research Service. International Economics Division and National Economics Division. *Government Intervention in Agriculture: Measurement, Evaluation, and Implications for Trade Negotiations.* Foreign Agricultural Economic Report No. 229. Washington, D.C.: U.S. Government Printing Office, April 1987.

United States. Department of Agriculture. Foreign Agricultural Service. *Foreign Agriculture* 12, No. 29 (July 15, 1974).

——. "Japan Institutes New Trade Measures, Cuts Tariffs on Some Agricultural Products." *Agricultural Trade Policy*, FAS Circular ATP-7-72, December 1972.

——. *"Nontariff Barriers Affecting Trade in Agricultural Products—Canada."* Agricultural Trade Policy, FAS Circular ATP–3–72, July 1972.

——. "Nontariff Barriers Affecting Trade in Agricultural Products—Denmark." *Agricultural Trade Policy,* FAS Circular ATP–5–72, September 1972.

——. "Nontariff Barriers Affecting Trade in Agricultural Products—EEC." *Agricultural Trade Policy,* FAS Circular ATP–10–72, June 1972.

——. "Nontariff Barriers Affecting Trade in Agricultural Products—Ireland." *Agricultural Trade Policy,* FAS Circular ATP–9–72, December 1972.

——. "Nontariff Barriers Affecting Trade in Agricultural' Products—Japan." *Agricultural Trade Policy,* FAS Circular ATP–2–72, June 1972.

——. "Nontariff Barriers Affecting Trade in Agricultural Products—Norway." *Agricultural Trade Policy,* FAS Circular ATP–4–72, September 1972.

——. "Nontariff Barriers Affecting Trade in Agricultural Products—Spain." *Agricultural Trade Policy,* FAS Circular ATP–6–72, September 1972.

——. "Nontariff Barriers Affecting Trade in Agricultural Products—Sweden." *Agricultural Trade Policy,* FAS Circular ATP–1–72, March 1972.

——. "Nontariff Barriers Affecting Trade in Agricultural Products—United Kingdom." *Agricultural Trade Policy,* FAS Circular ATP–8–72, December 1972.

——. *Trade Policies and Market Opportunities for U.S. Farm Exports: 1988 Annual Report.* Washington, D.C.: U.S. Government Printing Office, 1989.

——. *World Agricultural Production and Trade,* Washington, D.C., December 1973.

United States. Department of Commerce. *International Commerce.* Washington, D.C.: U.S. Government Printing Office, 1971.

United States. Senate. *Nontariff Trade Barriers.* Report to the Committee on Finance. T.C. Publication No. 665. Washington, D.C., 1974.

United States Statutes at Large. 88th Cong., 2^{d} sess., 1964. Vol. 78. Washington, D.C.: U.S. Government Printing Office, 1965.

United States. Tariff Commission. *Nontariff Trade Barriers.* T.C. Publication No. 665, April 1974.

———. *Summaries of Trade and Tariff Information. Animals and Meat.* Vol. 1, T.C. Publication No. 250. Washington, D.C.: U.S. Government Printing Office, 1968.

Vollrath, Thomas L. *Competitiveness and Protection in World Agriculture*, Agriculture Info. Bulletin No. 567. Washington, D.C.: Economic Research Service, USDA, July 1989.

Webb, Alan J., Michael Lopez, and Renata Penn. *Estimates of Producer and Consumer Subsidy Equivalents: Government Intervention in Agriculture*, 1982–87, Statistical Bulletin No. 803. Washington, D.C.: Economic Research Service, USDA, April 1990.

Yates, P.L. *Food, Land and Manpower in Western Europe.* London: Macmillan, 1960.

———. *Forty Years of Foreign Trade.* London: George Allen and Unwin, 1959.

Zietz, Joachim, and Alberto Valdés. *Agriculture in the GATT: An Analysis of Alternative Approaches to Reform.* Research Report 70. Washington, D.C.: International Food Policy Research Institute, November 1988.

Index

191